Cambridge Primary

ENDORSED BY

CAMBRIDGE
International Examinations

D1797262

Ready to Go Lessons for Maths

Step-by-step
lesson plans for
Cambridge Primary

Stage 3

Helen Lewis

Series editor: Paul Broadbent

HODDER
EDUCATION
AN HACHETTE UK COMPANY

Every effort has been made to trace all copyright holders, but if any have been inadvertently overlooked the Publishers will be pleased to make the necessary arrangements at the first opportunity.

Although every effort has been made to ensure that website addresses are correct at time of going to press, Hodder Education cannot be held responsible for the content of any website mentioned in this book. It is sometimes possible to find a relocated web page by typing in the address of the home page for a website in the URL window of your browser. Websites included in this text have not been reviewed as part of the Cambridge endorsement process.

Hachette UK's policy is to use papers that are natural, renewable and recyclable products and made from wood grown in sustainable forests. The logging and manufacturing processes are expected to conform to the environmental regulations of the country of origin.

Orders: please contact Bookpoint Ltd, 130 Milton Park, Abingdon, Oxon OX14 4SB. Telephone: (44) 01235 827720. Fax: (44) 01235 400454. Lines are open 9.00–5.00, Monday to Saturday, with a 24-hour message answering service. Visit our website at www.hoddereducation.com.

© Helen Lewis 2013
First published in 2013 by
Hodder Education,
An Hachette UK Company
338 Euston Road
London NW1 3BH

Impression number   5 4 3 2 1
Year   2017 2016 2015 2014 2013

Cover illustration by Peter Lubach
Illustrations by Planman Technologies
Typeset in ITC Stone Serif 10/12.5 by Planman Technologies
Printed in Great Britain by CPI Group (UK) Ltd, Croydon, CR0 4YY

A catalogue record for this title is available from the British Library.

ISBN: 978 1444 177589

Contents

Introduction   4

Overview chart   6

Term 1

Unit 1A: **Number and problem solving**   8
          Unit assessment   33

Unit 1B: **Geometry and problem solving**   35
          Unit assessment   49

Unit 1C: **Measure and problem solving**   51
          Unit assessment   67

Term 2

Unit 2A: **Number and problem solving**   69
          Unit assessment   95

Unit 2B: **Measure and problem solving**   97
          Unit assessment   113

Unit 2C: **Handling data and problem solving**   115
          Unit assessment   127

Term 3

Unit 3A: **Number and problem solving**   129
          Unit assessment   157

Unit 3B: **Geometry and problem solving**   159
          Unit assessment   173

Unit 3C: **Measure and problem solving**   175
          Unit assessment   191

Introduction

About the series

Ready to Go Lessons for Maths is a series of photocopiable resource books providing creative teaching strategies for primary teachers. These books support the revised Cambridge Primary curriculum frameworks for English, Mathematics and Science at Stages 1–6 (ages 5–11). They have been written by experienced primary teachers to reflect the different teaching approaches recommended in the Cambridge Primary Teacher Guides. The books contain lesson plans and photocopiable support materials, with a wide range of activities and appropriate ideas for assessment and differentiation. As the books are intended for international schools we have taken care to ensure that they are culturally sensitive.

Cambridge Primary

The Cambridge Primary curriculum frameworks show schools how to develop the learners' knowledge, skills and understanding in English, Mathematics and Science. They provide a secure foundation in preparation for the Cambridge Secondary 1 (lower secondary) curriculum. The ideas in this book can also be easily incorporated into existing curriculum frameworks already in your school.

How to use this book

This book covers each of the units of the scheme of work for Mathematics at Stage 3. It can be worked through systematically (as all the learning objectives are covered), or used to support areas where you feel you need more ideas. It is not prescriptive – it gives ideas and suggestions for you to incorporate into your own teaching as you see fit.

Each step-by-step lesson plan shows you the learning objectives you will cover, the resources you will need and how to deliver the lesson. Each lesson includes a Mental / oral starter activity, Main activities and a Plenary that draws the lesson to a close and recaps the learning objectives. Success criteria are provided in the form of questions to help you assess the learners' level of understanding. The 'Differentiation' section provides support for the less-able learners and extension ideas for the more able.

For each lesson plan there is at least one supporting photocopiable activity page. At the end of each unit there are also suggestions for assessment activities. Answers to activities can be found at www.hoddereducation.com/checkpointextras.

Learning objectives

The *Mathematics Curriculum Framework* provides a set of learning objectives for each stage. At the start of each lesson you need to re-phrase the learning objectives into child-friendly language so that you can share them with the learners at the outset. It sometimes helps to express them as *We are learning to / about …* statements. This really does help the learners to focus on the lesson's outcomes, for example: 'Understand what each digit represents in three-digit numbers and partition into hundreds, tens and units.' (Stage 3) could be introduced to the learners at the start of the lesson as: *We are learning about the value of each of the digits in a three-digit number.* To avoid unnecessary repetition we have not included such statements at the start of each lesson plan but it is understood that the teacher would do this.

The overview chart on pages 6–7 shows you how the learning objectives are covered in the lessons in this book.

Success criteria

These are the measures that the teacher and, eventually, the learner will be able to use to assess the outcome of the learning that has taken place in each lesson. They are included as a series of questions, which will help you as teacher to assess the learners' understanding of the skills and knowledge covered in the lesson.

Problem-solving skills

Maths teaching is concerned with more than just the learning of mathematical facts. Problem-solving skills are also **essential** and are planned as an on-going and sequential part of each unit.

The activities in this book will show you how to incorporate the practical nature of problem-solving so that it is part of the teaching process. Problem-solving objectives are worked into every teaching unit, with these skills underpinning all other strands to help the learners understand mathematical relationships and functions. These skills need to be used regularly in familiar and new contexts in order for the learners to become mathematical thinkers who are capable of questioning, reasoning and seeking answers through investigation.

The key to successful mathematical problem-solving teaching lies in providing the learners with opportunities to learn by doing, that is, through **active learning**.

Mental / oral starters

These are an important part of each lesson, consisting of whole-class, teacher-led interactive activities. The purpose varies for each lesson, and can include:

- practice and consolidation of existing skills – often mental calculation but also properties of shape and the language of number
- quick recall – to secure knowledge of number facts and build up speed and accuracy
- revisiting previous learning – to return to aspects of Maths that may have caused difficulty or to strengthen the learners' knowledge and use of mental or written strategies
- preparation for main part of the lesson – linked to the objectives for the lesson to support the learning.

Formative assessment

Formative assessment is on-going assessment that occurs in every lesson and informs the teacher and learners of the progress they are making, linked to the success criteria. The types of questions to ask that will support teachers in making formative assessments have been incorporated into each lesson in the 'Success criteria' sections.

One of the advantages of formative assessment is that any problems that arise during the lesson can be responded to immediately. Formative assessment influences the next steps in learning and may influence changes in planning and / or delivery for subsequent lessons.

Summative assessment

Summative assessment is essential at the end of each unit of work to assess exactly what the learners know, understand and can do. The assessment sections at the end of each unit are designed to provide you with a variety of opportunities to check the learners' understanding of the unit. These activities can include specific questions for teachers to ask, activities for the learners to carry out (independently, in pairs or in groups) or written assessment.

The information gained from both the formative and summative assessment ideas can then be used to inform future planning in order to close any gaps in the learners' understanding as recommended by *Assessment for Learning* (AFL).

Appropriate use of ICT

At the planning stage teachers need to consider how the use of ICT in a lesson will enhance the learning process. Ensure that the ICT resources you use support and promote the learners' understanding of the learning objectives. Activities included in this book have been designed to be carried out without the need for state-of-the-art ICT facilities. Suggestions have also been included for schools with internet access and / or the use of interactive whiteboards. This is in order to cater for most teachers' needs.

In these lessons the author sometimes asks for the teacher to display an enlarged version of the photocopiable page at the front of the class. We have not specified whether this should be using an overhead projector, interactive whiteboard or flipchart, as schools will have different resources available to them.

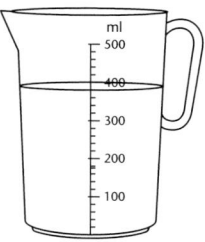

We hope that using these resources will give you confidence and creative ideas in delivering the Cambridge Primary curriculum framework.

Paul Broadbent, Series Editor

Overview chart

		Lesson	Framework codes	Page
Term 1	**Unit 1A: Number and problem solving**	Numbers to 200 and beyond	3Nn1 3Nn10	8
		Numbers to 1000	3Nn2 3Nn9	10
		Place value 1	3Nn3 3Nn5	13
		Place value 2	3Nn6 3Nc20	15
		Number patterns	3Nn4 3Nc5 3Ps5	17
		Addition and subtraction facts 1	3Nc1 3Nc11	19
		Addition and subtraction facts 2	3Nc2 3Pt5 3Ps6	21
		Addition strategies 1	3Nc12 3Nc16 3Pt4	23
		Addition strategies 2	3Nc9 3Nc10 3Pt1	25
		Multiplication and division 1	3Nc3 3Nc25 3Ps6	27
		Multiplication and division 2	3Nc4 3Pt3 3Pt12 3Ps1 3Ps3	29
		Doubling and halving 1	3Nc19 3Pt1 3Ps2 3Ps6	31
	Unit assessment			33
	Unit 1B: Geometry and problem solving	Co-ordinates 1	3Gp1	35
		Symmetry 1	3Gs5 3Ps7	37
		Angles 1	3Gs2 3Gs8	39
		2D shapes 1	3Gs1 3Gs2 3Pt8	41
		2D shapes 2	3Gs1 3Gp3 3Pt8	43
		3D shapes 1	3Gs3 3Gs4 3Gs6	45
		3D shapes 2	3Gs3 3Gs7 3Pt9	47
	Unit assessment			49
	Unit 1C: Measure and problem solving	Money 1	3Mm1 3Mm2	51
		Money 2	3Pt1 3Ps1	53
		Length 1	3Ml1 3Ml4 3Pt2	55
		Mass 1	3Ml2 3Ml3 3Ml5	57
		Capacity 1	3Ml3 3Pt2 3Pt11 3Pt12	59
		Time 1	3Ml1 3Mt1 3Pt2	61
		Time 2	3Mt2	63
		Time 3	3Ml5 3Ps2	65
	Unit assessment			67

		Lesson	Framework codes	Page
Term 2	**Unit 2A: Number and problem solving**	Comparing and ordering numbers 1	3Nn9 3Nn11 3Nn12	69
		Place value 3	3Nn3 3Nn5 3Nn6 3Nc18	71
		Place value 4	3Nn7 3Ps8	73
		Estimating and rounding 1	3Nn8 3Nn13	75
		Estimating and rounding 2	3Pt3 3Pt10 3Pt11 3Ps2 3Ps9	77
		Addition and subtraction facts 3	3Nn4 3Nc2 3Ps3	79
		Addition and subtraction strategies 1	3Nc14 3Pt1 3Pt4 3Pt5 3Pt12	81
		Addition and subtraction strategies 2	3Nc15 3Nc17 3Ps1	83
		Multiplication and division facts 1	3Nc3 3Nc26 3Pt6 3Ps5	85
		Multiplication and division facts 2	3Nc4 3Ps6	87
		Doubling and halving 2	3Nc6 3Nc7 3Nc19	89
		Multiplication and division strategies 1	3Nc21 3Pt1	91
		Multiplication and division strategies 2	3Nc24 3Pt3 3Pt7	93
	Unit assessment			95

	Lesson	Framework codes	Page
Unit 2B: Measure and problem solving	Money 3	3Mm1 3Mm2 3Pt11	97
	Money 4	3Ml5 3Pt1 3Pt12 3Ps1	99
	Length 2	3Ml3 3Ml4 3Ps2 3Ps4	101
	Time 4	3Ml5 3Mt1 3Mt4	103
	Time 5	3Mt2	105
	Time 6	3Ml5 3Mt3 3Pt10	107
	Mass 2	3Ml1 3Ml3 3Ml5 3Pt2	109
	Capacity 2	3Ml1 3Ml2 3Ml3 3Ml5	111
Unit assessment			113
Unit 2C: Handling data and problem solving	Tally charts and frequency tables	3Dh2 3Ps4	115
	Pictograms	3Dh2	117
	Bar charts	3Dh2	119
	Carroll diagrams	3Dh3	121
	Venn diagrams	3Dh3	123
	Data handling project	3Dh1 3Dh2	125
Unit assessment			127

	Lesson	Framework codes	Page
Unit 3A: Number and problem solving	Place value 5	3Nc18 3Ps5 3Ps8	129
	Comparing and ordering numbers 2	3Nn7 3Nn11 3Nn12	131
	Estimating and rounding 3	3Nn8 3Nn13 3Pt10 3Pt11	133
	Addition and subtraction facts 4	3Nc2	135
	Addition and subtraction strategies 3	3Nc14 3Pt1 3Pt4 3Pt5 3Ps2	137
	Addition and subtraction strategies 4	3Nc15 3Nc17 3Ps3 3Ps9	139
	Doubling and halving 3	3Nc6 3Nc7 3Nc8 3Ps6	141
	Multiplication strategies	3Nc3 3Nc22 3Pt6	143
	Division strategies	3Nc23 3Nc24 3Nc26 3Pt3 3Pt7	145
	Fractions 1	3Nn14 3Ps1	147
	Fractions 2	3Nn15 3Nn16	149
	Fractions 3	3Nn17 3Nn18	152
	Fractions 4	3Nn19 3Nn20 3Nc4 3Nc21	154
Unit assessment			157
Unit 3B: Geometry and problem solving	2D shapes 3	3Gs1 3Gs2	159
	2D shapes 4	3Pt8	161
	Symmetry 2	3Gs5 3Ps7	163
	Angles 2	3Gs8 3Gp3 3Gp4	165
	Co-ordinates 2	3Gp1 3Gp2	167
	3D shapes 3	3Gs4 3Gs6 3Pt9	169
	3D shapes 4	3Gs3 3Gs7	171
Unit assessment			173
Unit 3C: Measure and problem solving	Time 7	3Ml5 3Mt1 3Mt4	175
	Time 8	3Mt2	177
	Time 9	3Ml5 3Mt3 3Pt10	179
	Money 5	3Mm1 3Mm2 3Ps2	181
	Money 6	3Ml5 3Pt1 3Pt12 3Ps1	183
	Length 3	3Ml1 3Ml2 3Ml4	185
	Mass 3	3Ml1 3Ml3 3Pt2	187
	Capacity 3	3Ml2 3Ml5	189
Unit assessment			191

Term 3

Unit 1A: Number and problem solving

Numbers to 200 and beyond

Starter

- Before the lesson make a counting stick by cutting a 1-metre length of wooden dowelling and painting every 10 cm in alternate colours.

- To use the counting stick, hold it horizontally, and point to the left end as viewed by the learners (your right). Point to each division in turn, counting aloud as you go. To count down, start at the right end of the stick (your left).
- Display an enlarged copy of the hundred square on photocopiable page 9, and use it to count in ones from 101 to 200 and back.
- Using the counting stick:
  - count on and back in ones from various numbers between 101 and 200; include one sequence that extends below 101 and one that extends past 200
  - point to one end and say a number, for example: *This is 153*; point to one of the other lines on the counting stick and ask: *What number is this?*

Main activities

- Display an enlarged copy of the number line made from photocopiable page 9. Count on and back in tens.
- Organise the learners into pairs. Give each pair a copy of the number line and a six-sided dice. Ask the learners to point to given multiples of 10 on their number lines (for example 170, 520, 690, 360, and so on).

- Demonstrate rolling a six-sided dice three times, using the numbers (in any order) to make a three-digit number, and marking this number on the number line.
- Ask the learners to play a game of 'Target 450' in pairs: Players take it in turns to generate a three-digit number by rolling the dice three times. They mark this number on the number line and write their initials next to it. The game ends as soon as both players have made ten numbers. The winner is the player who has made the number closest to 450.

Plenary

- Write the following digits on the board: 2, 4, 7.
- Ask the learners which three-digit number they would choose to make from those digits if they were playing a game of 'Target 450', and why.
- Explore ideas on a number line (for example 427 is 23 away, but 472 is only 22 away).

The second hundred

101	102	103	104	105	106	107	108	109	110
111	112	113	114	115	116	117	118	119	120
121	122	123	124	125	126	127	128	129	130
131	132	133	134	135	136	137	138	139	140
141	142	143	144	145	146	147	148	149	150
151	152	153	154	155	156	157	158	159	160
161	162	163	164	165	166	167	168	169	170
171	172	173	174	175	176	177	178	179	180
181	182	183	184	185	186	187	188	189	190
191	192	193	194	195	196	197	198	199	200

Number line

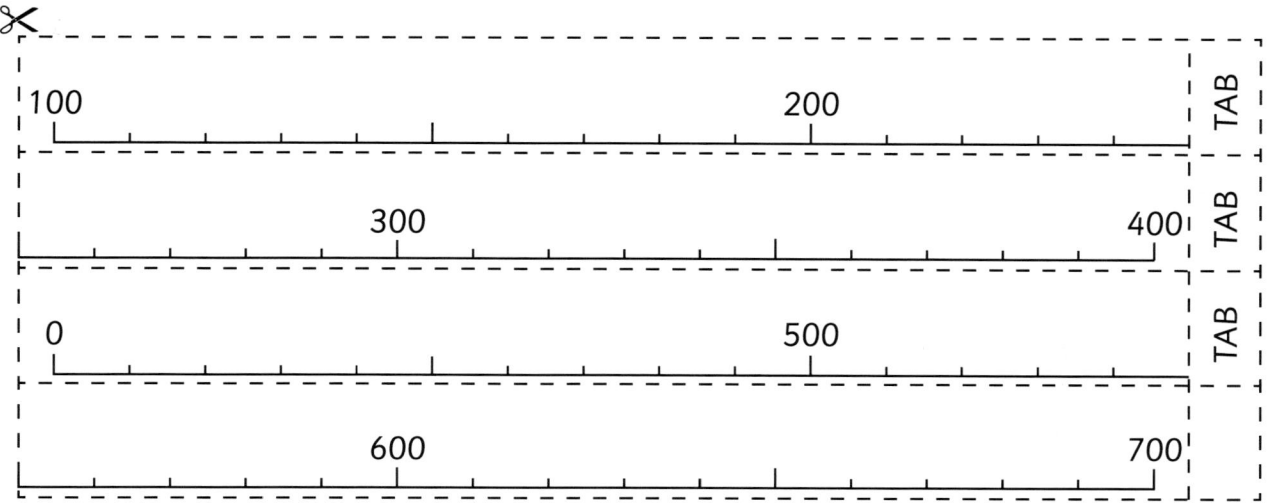

Numbers to 1000

Learning objectives

- Read and write numbers to at least 1000. (3Nn2)
- Place a three-digit number on a number line marked in multiples of 100. (3Nn9)

Resources

Number cards from photocopiable pages 11 and 12 (enlarged or copied onto acetate); pencils; plain paper; 0 to 9 ten-sided dice (see template on photocopiable page 72) or sets of 0 to 9 digit cards; A3 paper; rulers.

Starter

- Before the lesson enlarge photocopiable pages 11 and 12 onto A3 card and cut out.
- Display the number cards one at a time. Ask the learners to say the numbers out loud. Gradually increase the pace.
- Give each learner paper and a pencil. Read aloud a number from a card, without showing the card to the learners. Ask the learners to write down the number in figures. Reveal the card to show the correct answer. Gradually increase the pace.

Main activities

- Ask the learners to find all the three-digit numbers that can be made with even digits only.
- Discuss strategies used to find all the possible numbers (for example using knowledge of place value to work out the smallest possible number that can be made, and then working systematically to find the next largest number, and so on). Ask: *Which of these numbers is closest to 500?*
- Teach the following game: organise the learners into pairs and give each pair a sheet of A3 paper and either a ten-sided dice or a set of 0 to 9 digit cards.

- On the board draw a number line to 1000, labelling zero and the multiples of 100. Make 500 stand out in some way and ask the learners to copy the number line onto their A3 sheet.
- Ask pairs to take it in turns, using either the ten-sided dice or the 0 to 9 digit cards, to generate a three-digit number made with even digits only. They should mark this number on the number line and write their initials next to it. The game ends as soon as both players have made ten numbers. The winner is the player who has made the number closest to 500.

Plenary

- Ask the learners to explain the strategies they used in the game.
- *What would be the best three digits you could possibly throw? Why?*

Success criteria

Ask the learners:

- How do you write eight hundred and four? Five hundred and sixty? Three hundred and nineteen?
- What is the same and what is different about the numbers 654 and 456?

Ideas for differentiation

Support: In the second starter activity, pair these learners with a more-able partner.

Extension: Play a game of 'Target 1000' with these learners: Extend the number line to 2000. Make two three-digit numbers, then add them to make a number as close as possible to 1000.

Number cards 1

345	323	699
621	237	485
716	986	1000
812	164	468
444	708	179
1001	790	503
245	535	1010
951	607	872

Number cards 2

534	1025	602
377	247	818
722	968	1019
161	554	759
496	614	103
1040	536	289
693	581	475
940	725	830

Place value 1

- Count on and back in ones, tens and hundreds from two- and three-digit numbers. (3Nn3)
- Understand what each digit represents in three-digit numbers and partition into hundreds, tens and units. (3Nn5)

Photocopiable page 14; large display, and tabletop, place value cards (showing hundreds, tens and units); pencils; plain paper.

Starter

- Display an A3 copy of photocopiable page 14. From a variety of two- and three-digit starting numbers, count on and back first in ones, then in tens and then in hundreds.
- When counting in tens and hundreds ask the learners to describe any patterns they can see in the numbers. (When counting on and back in tens the units digits stay the same, and the tens digits get one more or one less each time; when counting in hundreds, the units and tens digits stay the same and the hundreds digits get one more or one less each time.)
- Encourage the learners to continue counting past 600.
- Finally, challenge the learners to count with their eyes closed, so that they cannot see the hundred squares.

Main activities

- Make a three-digit number using the large display place value cards (for example 426).

- Point to one of the digits and ask: *What is the value of this digit?* (for example the 2 in 426 has a value of 20). Confirm whether answers are correct by pulling the place value cards apart. Repeat for other three-digit numbers.
- Organise the learners into pairs and give each pair a set of place value cards and paper and pencils. Show three place value cards (for example 60, 700, 1). Ask: *What number do these cards make?* Ask the pairs to make the number using place value cards and write an addition (for example 700 + 60 + 1 = 761).
- Combine pairs to form groups of four and ask pairs to take it in turns to make a secret three-digit number using the place value cards. The other pair must guess the number by asking questions about the digits in the number, for example: *Does your number contain a six?* (Yes.) *Is its value sixty?* (No.)

Plenary

- Ask volunteers to make three-digit numbers using the large display place value cards.
- Ask: *Which digit is worth the most / least in this number? How do you know?*

Ask the learners:

- (Pointing to a digit in a three-digit number made on place value cards:) What is this digit worth? How could you check your answer?
- Make a three-digit number using the place value cards. Read the number aloud. Explain what each digit is worth.

Support: In the paired main activity, support these learners by grouping them together and working with them to devise a simple game.

Extension: In the paired main activity, challenge these learners to partition the numbers without using the place value cards.

Hundred squares

1	2	3	4	5	6	7	8	9	10
11	12	13	14	15	16	17	18	19	20
21	22	23	24	25	26	27	28	29	30
31	32	33	34	35	36	37	38	39	40
41	42	43	44	45	46	47	48	49	50
51	52	53	54	55	56	57	58	59	60
61	62	63	64	65	66	67	68	69	70
71	72	73	74	75	76	77	78	79	80
81	82	83	84	85	86	87	88	89	90
91	92	93	94	95	96	97	98	99	100

101	102	103	104	105	106	107	108	109	110
111	112	113	114	115	116	117	118	119	120
121	122	123	124	125	126	127	128	129	130
131	132	133	134	135	136	137	138	139	140
141	142	143	144	145	146	147	148	149	150
151	152	153	154	155	156	157	158	159	160
161	162	163	164	165	166	167	168	169	170
171	172	173	174	175	176	177	178	179	180
181	182	183	184	185	186	187	188	189	190
191	192	193	194	195	196	197	198	199	200

201	202	203	204	205	206	207	208	209	210
211	212	213	214	215	216	217	218	219	220
221	222	223	224	225	226	227	228	229	230
231	232	233	234	235	236	237	238	239	240
241	242	243	244	245	246	247	248	249	250
251	252	253	254	255	256	257	258	259	260
261	262	263	264	265	266	267	268	269	270
271	272	273	274	275	276	277	278	279	280
281	282	283	284	285	286	287	288	289	290
291	292	293	294	295	296	297	298	299	300

301	302	303	304	305	306	307	308	309	310
311	312	313	314	315	316	317	318	319	320
321	322	323	324	325	326	327	328	329	330
331	332	333	334	335	336	337	338	339	340
341	342	343	344	345	346	347	348	349	350
351	352	353	354	355	356	357	358	359	360
361	362	363	364	365	366	367	368	369	370
371	372	373	374	375	376	377	378	379	380
381	382	383	384	385	386	387	388	389	390
391	392	393	394	395	396	397	398	399	400

401	402	403	404	405	406	407	408	409	410
411	412	413	414	415	416	417	418	419	420
421	422	423	424	425	426	427	428	429	430
431	432	433	434	435	436	437	438	439	440
441	442	443	444	445	446	447	448	449	450
451	452	453	454	455	456	457	458	459	460
461	462	463	464	465	466	467	468	469	470
471	472	473	474	475	476	477	478	479	480
481	482	483	484	485	486	487	488	489	490
491	492	493	494	495	496	497	498	499	500

501	502	503	504	505	506	507	508	509	510
511	512	513	514	515	516	517	518	519	520
521	522	523	524	525	526	527	528	529	530
531	532	533	534	535	536	537	538	539	540
541	542	543	544	545	546	547	548	549	550
551	552	553	554	555	556	557	558	559	560
561	562	563	564	565	566	567	568	569	570
571	572	573	574	575	576	577	578	579	580
581	582	583	584	585	586	587	588	589	590
591	592	593	594	595	596	597	598	599	600

Place value 2

● Find 1, 10, 100 more / less than two- and three-digit numbers. (3Nn6)

● Understand the effect of multiplying two-digit numbers by 10. (3Nc20)

Photocopiable pages 14 and 16; writing materials; standard six-sided dice.

Starter

- Display an A3 copy of photocopiable page 14. Point to a number. Ask: *What number is this? What is one more?* Repeat for other numbers. Ask: *How do you move on the hundred square to find one more?* (Move one space to the right.)

- Repeat for finding 10 and then 100 more. Ask the learners to suggest how to find 1, 10 and 100 less on the hundred square and test out their suggestions.

- Organise the learners into pairs and give each pair photocopiable page 14 and some writing materials. Ask pairs to find a given number and then write the number that is 1, 10 or 100 more or less, for example: *Find 516. Write the number that is 10 less.*

Main activities

- Display an A3 copy of photocopiable page 16. Explain how the function machine works. (A number goes in where it says 'In', the machine performs a calculation on the number, and the answer comes out of the machine where it says 'Out'.)

- Ask: *What calculation does this function machine perform?* (Multiplying by 10.) Ask: *If I put 15 into the machine, what number will come out?* (150.) *How did you work it out?* (For example wrote a zero on the end; moved the digits one place to the left; I know that 10 times 10 is 100 and 5 times 10 is 50 and I added them together.) Ask the learners to predict the function machine's output for a variety of two-digit input numbers.

- Working in pairs, ask the learners to roll a six-sided dice twice to make a two-digit number (for example 6 and 3 to make 36). Ask them to multiply the number by 10, and then mark the product, or the result of the multiplication (for example 360), on photocopiable page 14. The learners could make this into a game by choosing a target number to try to get closest to.

Plenary

- Write and display a three-digit multiple of 10 (for example 870). Say: *This number came out of the function machine. What number was put into the machine?*

- Repeat for other multiples of 10, including at least one multiple of 100 and at least one number above 1000.

Ask the learners:

● Find four hundred and nine on a hundred square. What number is 100 more? 10 more? 10 less?

● What happens when you multiply a whole number by 10?

Support: In the second main activity, support these learners by grouping them together and working with them.

Extension: In the second starter activity, encourage these learners to work without the hundred squares.

Function machine

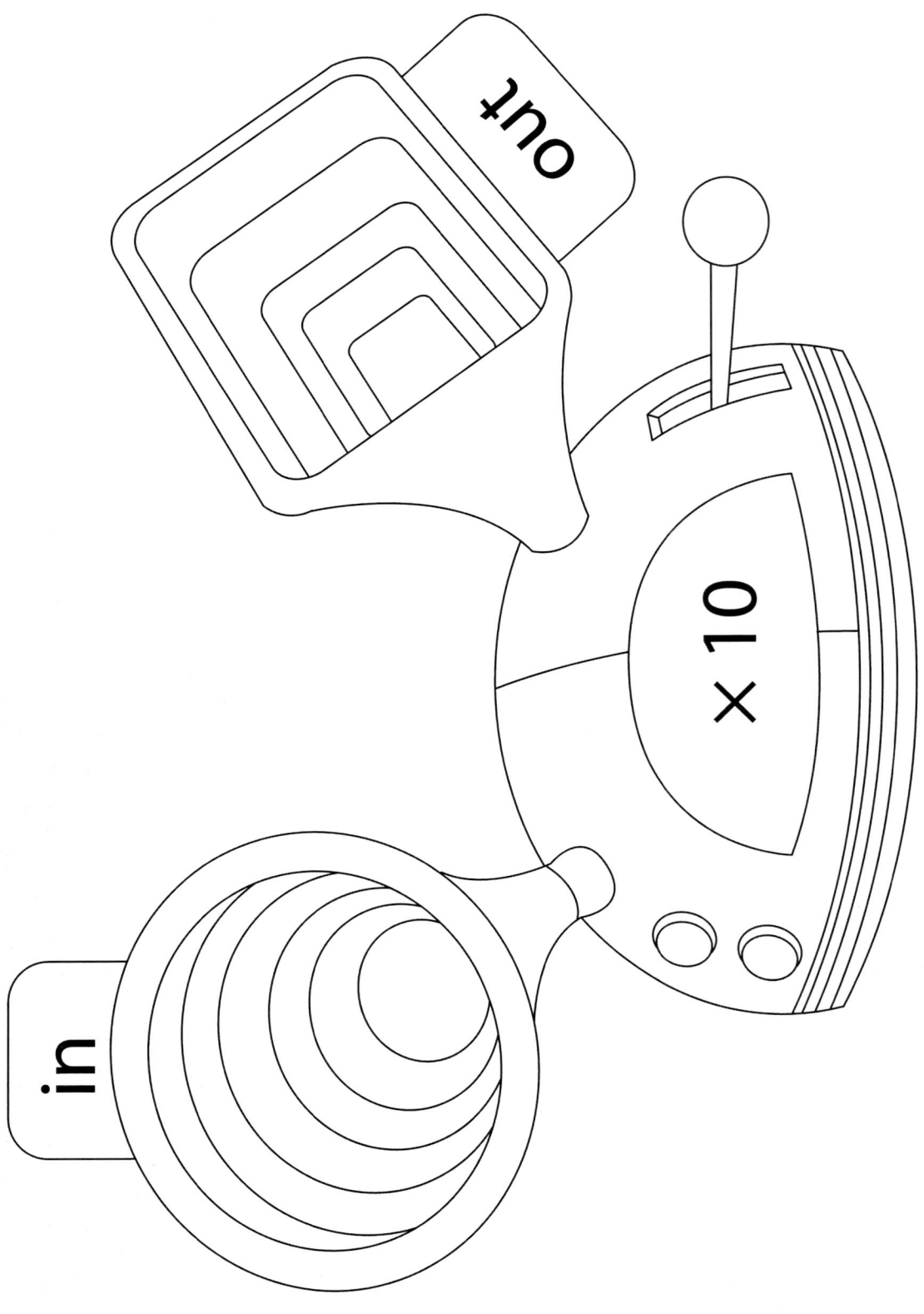

out

× 10

in

Number patterns

- Count on and back in steps of 2, 3, 4 and 5 to at least 50. (3Nn4)
- Recognise two- and three-digit multiples of 2, 5 and 10. (3Nc5)
- Describe and continue patterns which count on or back in steps of 2, 3, 4, 5, 10 or 100. (3Ps5)

Photocopiable page 18; coloured pens or pencils; ten-sided dice marked 0 to 9 (see template on photocopiable page 72).

Starter

- Display a copy of photocopiable page 18. On the first hundred square, count on from 2 in steps of 2 to at least 50, and then count back again. Ask the learners to describe any patterns they notice in the numbers. (These could be patterns in the position of the numbers on the grid, or patterns in the numbers' digits.) Repeat the activity, counting from 5 in steps of 5 and then from 10 in steps of 10.
- Challenge the learners to count on and back in 2s, 3s, 4s, 5s and 10s to at least 50 without looking at the hundred square.

Main activities

- Revise the term 'multiple' (a number that is made by multiplying together two whole numbers). Ask the learners to name a multiple of 10. Write a few suggestions on the board. For each number given, ask: *What number did you multiply by 10 to make that multiple of 10?*
- On the displayed copy of photocopiable page 18, mark the first few multiples of 10 by circling them in one colour. Mark the first few multiples of 5 and 2 in two other colours. Talk about the fact that some numbers are multiples of more than one of the numbers (for example 10 is a multiple of all three numbers).

- Give each learner a copy of photocopiable page 18 and ask them to copy and complete the pattern you have started (up to 200).
- Organise the learners into groups of four for a game. Ask them to take it in turns to roll three ten-sided dice and attempt to make a number that is a multiple of 2, 5 or 10. The learners score one point for every correct number they make. If the number is a multiple of all three numbers, they score three points.

Plenary

- Ask: *Which sorts of numbers scored the most points in the game?* (Multiples of 10.) *Why?* (Because every multiple of 10 is also a multiple of both 2 and 5, because 2 times 5 equals 10.)
- Ask: *Which numbers should you avoid using as units digits in the game?* (1, 3, 7 and 9.) *Why?* (Because numbers ending in these digits cannot be multiples of 2, 5 or 10.)

Ask the learners:

- What is the pattern in these numbers: 120, 110, 100, 90, 80?
- Can you write the next three numbers in this number pattern: 45, 50, 55, …?
- Name a number that is a multiple of both 2 and 5. What other number is this number a multiple of?

Support: In the second main activity, pair these learners with a more confident friend who will be able to support them.

Extension: In the game, ask these learners to find three-digit multiples of 2, 5 and 10 without referring to the patterns they've marked on the hundred squares.

Name: _____

The first two hundred squares

1	2	3	4	5	6	7	8	9	10
11	12	13	14	15	16	17	18	19	20
21	22	23	24	25	26	27	28	29	30
31	32	33	34	35	36	37	38	39	40
41	42	43	44	45	46	47	48	49	50
51	52	53	54	55	56	57	58	59	60
61	62	63	64	65	66	67	68	69	70
71	72	73	74	75	76	77	78	79	80
81	82	83	84	85	86	87	88	89	90
91	92	93	94	95	96	97	98	99	100

101	102	103	104	105	106	107	108	109	110
111	112	113	114	115	116	117	118	119	120
121	122	123	124	125	126	127	128	129	130
131	132	133	134	135	136	137	138	139	140
141	142	143	144	145	146	147	148	149	150
151	152	153	154	155	156	157	158	159	160
161	162	163	164	165	166	167	168	169	170
171	172	173	174	175	176	177	178	179	180
181	182	183	184	185	186	187	188	189	190
191	192	193	194	195	196	197	198	199	200

Cambridge Primary: Ready to Go Lessons for Maths Stage 3 © Hodder & Stoughton Ltd 2013

Addition and subtraction facts 1

Learning objectives

- Know addition and subtraction facts for all numbers to 20. (3Nc1)
- Use the = sign to represent equality. (3Nc11)

Resources

0 to 9 number fans (photocopiable page 24); photocopiable page 20; 0 to 10 number cards; 5 to 15 number cards; 0 to 20 number lines; 10 to 20 number cards.

Starter

- Revise addition facts with totals up to 10, and the associated subtraction facts.
- Give each learner a 0 to 10 number fan. Give the learners additions with totals no greater than 10 and subtractions with starting numbers no greater than 10. Present some questions verbally, using a variety of question forms, for example: *What is four plus two? Add together six and three … Eight subtract four is …?* Then write some questions, for example 10 – 3 = ☐, ☐ = 7 – 4. Ask the learners to show the answer on their number fan. Keep the pace brisk.

Main activities

- Display a copy of photocopiable page 20. Work through the first one or two questions in each column.
- Give each learner photocopiable page 20. Ask the less-able and average learners to complete section A and the more-able learners to complete section B.
- Play this game: Organise the learners into ability groups of between four and eight. Give each group two sets of 5 to 15 number cards. Shuffle the cards together and place in a face-down pile. Turn over the top two cards. The first person to call out the correct total is given the two cards. The winner is the player with the most cards when there are no more cards left in the pile.

Plenary

- Ask a volunteer to write an addition with a total up to 10.
- Ask: *What subtraction fact can we work out from this addition fact?* Ask: *What other addition fact with a total between 11 and 20 can we work out from this addition fact?*

Success criteria

Ask the learners:

- Can you write down an addition fact you know?
- Can you write down a subtraction fact you know from knowing this addition fact?
- Can you write down another subtraction fact you know from knowing this subtraction fact?

Ideas for differentiation

Support: In the game, give these groups two sets of 1 to 10 number cards instead of 5 to 15 number cards, and give each learner a 0 to 20 number line for support.

Extension: In the game, give these groups two sets of 10 to 20 number cards instead of 5 to 15 number cards.

Name: _____

Facts to 20

Complete the number sentences by writing the correct number in each box.
The first question in each section has been done for you.

Section A

1. $3 + 5 = 8$, so $13 + 5 = \boxed{18}$

2. $1 + 4 = 5$, so $1 + 14 = \boxed{}$

3. $4 + 4 = 8$, so $14 + 4 = \boxed{}$

4. $4 + 2 = 6$, so $4 + 12 = \boxed{}$

5. $3 + 6 = 9$, so $13 + 6 = \boxed{}$

6. $7 - 4 = 3$, so $17 - 4 = \boxed{}$

7. $9 - 4 = 5$, so $19 - 4 = \boxed{}$

8. $8 - 7 = 1$, so $18 - 7 = \boxed{}$

9. $10 - 6 = 4$, so $20 - 6 = \boxed{}$

10. $9 - 7 = 2$, so $19 - 7 = \boxed{}$

Section B

1. We know that $14 + 5 = \boxed{19}$ because $\boxed{4} + \boxed{5} = \boxed{9}$

2. We know that $6 + 12 = \boxed{}$ because $\boxed{} + \boxed{} = \boxed{}$

3. We know that $16 + 4 = \boxed{}$ because $\boxed{} + \boxed{} = \boxed{}$

4. We know that $3 + 11 = \boxed{}$ because $\boxed{} + \boxed{} = \boxed{}$

5. We know that $15 + 3 = \boxed{}$ because $\boxed{} + \boxed{} = \boxed{}$

6. We know that $18 - 5 = \boxed{}$ because $\boxed{} - \boxed{} = \boxed{}$

7. We know that $15 - 5 = \boxed{}$ because $\boxed{} - \boxed{} = \boxed{}$

8. We know that $16 - 4 = \boxed{}$ because $\boxed{} - \boxed{} = \boxed{}$

9. We know that $19 - 4 = \boxed{}$ because $\boxed{} - \boxed{} = \boxed{}$

10. We know that $17 - 3 = \boxed{}$ because $\boxed{} - \boxed{} = \boxed{}$

Cambridge Primary: Ready to Go Lessons for Maths Stage 3 © Hodder & Stoughton Ltd 2013

Addition and subtraction facts 2

Learning objectives

- Know the following addition and subtraction facts: multiples of 100 with a total of 1000; multiples of 5 with a total of 100. (3Nc2)
- Check subtraction by adding the answer to the smaller number in the original calculation. (3Pt5)
- Identify simple relationships between numbers. (3Ps6)

Resources

Counting stick; photocopiable page 22; several sets of 20 'multiples of 5' number cards (5 to 95 with two 50s).

Starter

- Using a counting stick, count in 100s to 1000 and back.
- Point to a line on the counting stick. Ask: *What's this number? What number do you need to add to this to make one thousand?* Cover each multiple of 100 in a random order. Repeat for subtraction questions, for example: *What's this number? What number do you need to subtract from one thousand to make this number?*
- Ask the learners whether any of the number facts they know already could help them remember multiples of 100 that make 1000. (Single-digit numbers that total 10 or multiples of 10 that total 100.)

Main activities

- Display a copy of photocopiable page 22. Highlight a pair of numbers and derive a family of facts, for example 85 + 15 = 100; 15 + 85 = 100; 100 – 85 = 15; 100 – 15 = 85.

- Ask questions that can be answered from the diagram, for example: *What is one hundred subtract thirty-five?* (65.) *What number do you need to add to twenty to make one hundred?* (80.)
- Play this game: Organise the learners into fours. Give each group one set of number cards. Shuffle the cards and deal five to each player. Tell the players to take it in turns to play a card. Whoever has the 'partner card' (i.e. showing the complement to 100) shows it, and puts both cards face down in front of them. The winner is the person with the most cards when all the cards have been played.

Plenary

- Play this game: Call two volunteers to the front of the class. Say a multiple of 5 between 5 and 95. Tell the players to say the complement to 100 as quickly as they can. The player who answers first should carry on playing. Replace the other player with a new volunteer.

Success criteria

Ask the learners:

- (Pointing to a multiple of 5 or 100:) What number do you need to add to this number to make 100 / 1000?
- (Pointing to a multiple of 5 or 100:) What number do you need to subtract from 100 / 1000 to make this number?

Ideas for differentiation

Support: Provide these learners with photocopiable page 22 to use during the game.

Extension: After they have played the game once, ask these learners to devise their own game using the number cards.

Multiples of 5 that make 100

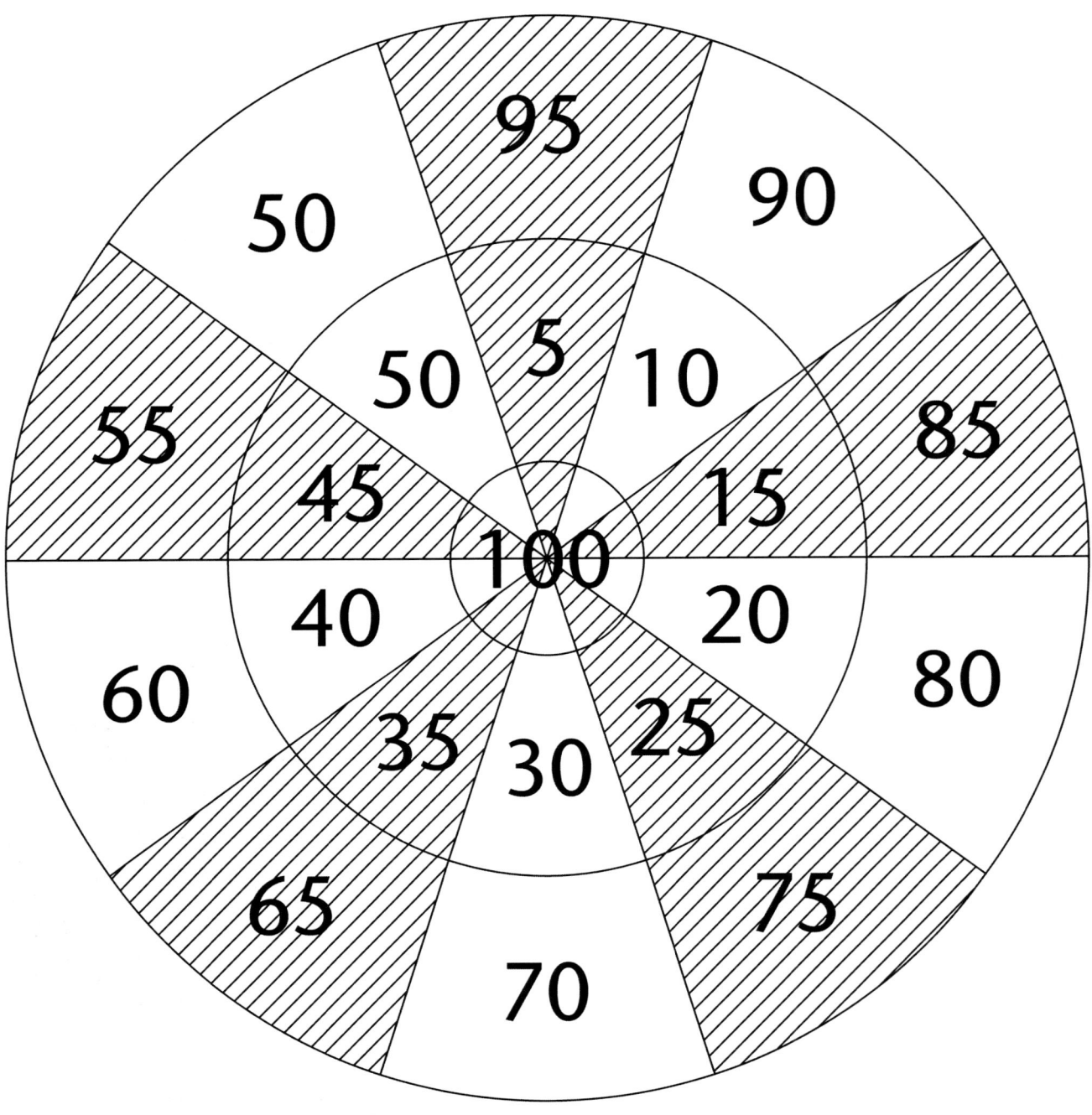

Addition strategies 1

Learning objectives

- Add several small numbers. (3Nc12)
- Re-order an addition to help with the calculation. (3Nc16)
- Check the results of adding two numbers using subtraction, and several numbers by adding in a different order. (3Pt4)

Resources

One large and lots of small 0 to 9 number fans (see photocopiable page 24); ten-sided dice marked 0 to 9 (see template on photocopiable page 72); calculators; six-sided dice (marked with the digits 1 to 6 rather than with spots); calculators.

Starter

- Before the lesson, make number fans using photocopiable page 24. Copy the learners' number fans onto A4 card. Enlarge the teacher's copy onto A3. Cut out the pieces and then laminate if possible. Punch a hole through each circle. Thread the pieces in order onto a book ring.
- Give each learner a small 0 to 9 number fan. Hold up two single-digit numbers on the large number fan. Ask the learners to show the total as quickly as they can on their number fan. Avoid making a total of 11, as this number can't be made on the number fan.
- Ask: *How did you work out the answer if you didn't 'just know it'?* Encourge the learners to describe their mental methods, for example identifying the larger number; keeping it in their head, and then counting on using fingers; using compensation from a known fact, such as 7 + 6 is one more than 6 + 6.
- Repeat at a brisker pace.

Main activities

- Write three single-digit numbers on the board in a triangular arrangement. Ask the learners to give the total.
- Ask: *How did you work out the answer?* Strategies the learners describe may include:

- ordering the numbers and adding them together in order, starting with the largest
- finding a pair of numbers whose total they already know and then adding the third number to the total by counting on.

Discuss the types of number pairs you might be on the lookout for (for example pairs of numbers that total 10, or doubles).

- Play this game: Organise the learners into pairs. Give each pair three ten-sided dice. Tell the learners to take it in turns to throw all three dice, calculate the total mentally and say the answer, then explain to their partner how they worked it out. The partner must check the answer by adding the numbers in a different order. The learners can devise their own scoring systems.

Plenary

- On the board write 12, 8, 8. Ask the learners to find the total.
- Discuss strategies used and how to check your answer.
- Ask the learners to work in pairs for a few minutes, making a list of real-life situations in which they might need to add these three numbers together. Share ideas.

Success criteria

Ask the learners:

- What is the total of these three numbers: 3, 6, 7?
- Explain how you worked out the answer.
- How could you check whether your answer is right?
- Check your answer. Was it right?

Ideas for differentiation

Support: In the game, give these learners six-sided dice (marked with the digits 1 to 6 rather than with spots). Provide the learners with calculators to check each other's answers.

Extension: In the game, challenge these learners to add four numbers together by throwing four ten-sided dice.

0 to 9 number fans

0 1 2 3

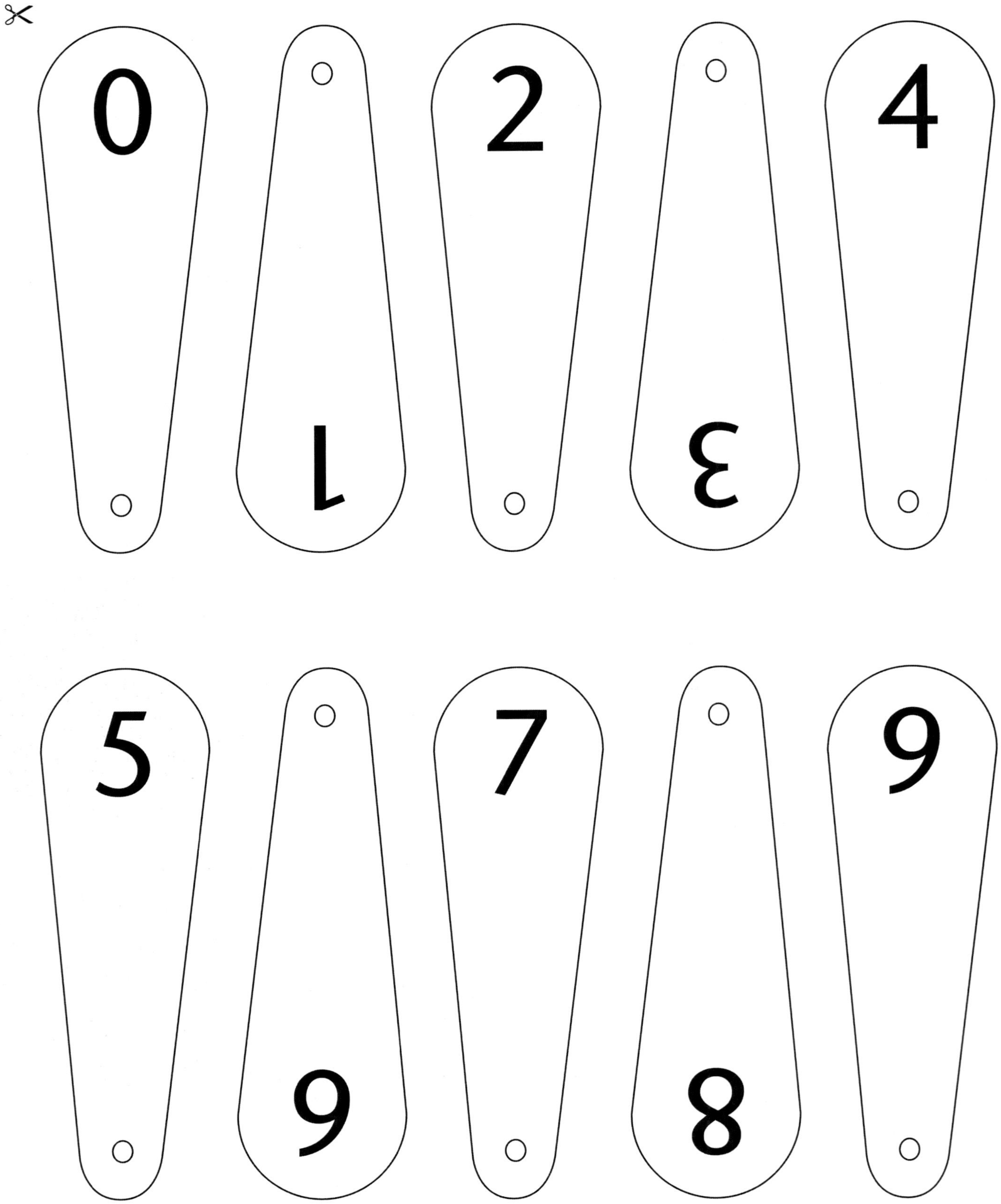

0

L

2

ε

4

5

9

7

8

9

Addition strategies 2

- Add and subtract 10 and multiples of 10 to and from two- and three-digit numbers. (3Nc9)
- Add 100 and multiples of 100 to three-digit numbers. (3Nc10)
- Choose appropriate mental strategies to carry out calculations. (3Pt1)

Resources

Large number cards showing the multiples of 10 to 100; large number cards showing the multiples of 100 to 1000; large and small place value cards (Th, H, T, U); photocopiable page 26; six-sided dice (marked with the digits 1 to 6 rather than with spots).

Starter

- Show two multiples-of-10 number cards, asking the learners to find the total. Start with totals within 100, and then extend to larger totals. Discuss methods of calculation, such as: starting at the larger number and then counting on in tens; using place value knowledge and known facts (3 + 8 = 11 so 30 + 80 = 110); using compensation starting from a known fact (30 + 70 = 100 so 30 + 80 is 10 more).
- Repeat the activity with multiples of 100. Start with totals within 1000, and extend to larger totals. Discuss methods, drawing comparisons with the methods for adding multiples of 10, for example starting at the larger number and then counting on in hundreds; using place value knowledge and known facts (3 + 8 = 11 so 300 + 800 = 1100); using compensation starting from a known fact (300 + 700 = 1000 so 300 + 800 is 100 more).

Main activities

- Organise the learners into pairs and give each pair a set of small place value cards.

- On the large place value cards make and display a two-digit number, such as 56. Ask the learners to add a multiple of 10 (for example 40) and make the answer with their place value cards (in this case, 96). Compare the starting number with the answer. Ask: *Which card is different?* (The tens card.) *Why?* (Because we're adding a whole number of tens.) Repeat for different pairs of numbers. Include some three-digit starting numbers.
- Repeat the activity making three-digit numbers and adding multiples of 100.
- Give each pair four dice and photocopiable page 26. Demonstrate what to do using an enlarged copy of photocopiable page 26. Ask the learners to take it in turns to roll the dice and write out the number sentence.

Plenary

- Divide the class into two teams. Choose two players of roughly equal ability, one from each team.
- Write out an addition, in the form ☐☐ + ☐0 or ☐☐☐ + ☐00. The first player to answer correctly wins their team a point.

Success criteria

Ask the learners:

- Write an addition with a three-digit number and a multiple of 100.
- What's the answer to the addition you wrote down?
- How did you work out the answer?

Ideas for differentiation

Support: In the final main activity, encourage these learners to use the place value cards for support.

Extension: Challenge these learners to do the final main activity without using the place value cards.

Name: _____

Addition strategies 2

Adding multiples of 10 to two-digit numbers

Roll three dice. Put one dice in each square. Write down the number sentence and its answer. Do this six times.

☐☐ + ☐0 =

1. _____

2. _____

3. _____

4. _____

5. _____

6. _____

Adding multiples of 100 to three-digit numbers

Roll four dice. Put one dice in each square. Write down the number sentence and its answer. Do this six times.

☐☐☐ + ☐00 =

1. _____

2. _____

3. _____

4. _____

5. _____

6. _____

Multiplication and division 1

Learning objectives

- Know multiplication / division facts for 2×, 3×, 5× and 10× tables. (3Nc3)
- Understand and apply the idea that multiplication is commutative. (3Nc25)
- Identify simple relationships between numbers. (3Ps6)

Resources

Photocopiable page 28; sticky notes; one large copy and plenty of small copies of a 1 to 100 hundred square; six-sided dice marked 2, 2, 5, 5, 10, 10; coloured counters; ten-sided dice marked 0 to 9 (see template on page 72); timers (optional).

Starter

- Display a copy of photocopiable page 28. Chant the times tables together.
- Use sticky notes to cover about a dozen of the products (answers) on the chart (about three in each times table). Point to a sticky note, asking the learners to call out the hidden product. Peel off the sticky note to reveal the answer.
- Select a fact on the times table chart. Ask the learners to derive three other facts from it, for example 6 × 3 = 18 also gives 3 × 6 = 18, 18 ÷ 3 = 6 and 18 ÷ 6 = 3.

Main activities

- Display a large hundred square. Point to a number and ask the learners whether it is a multiple of 2, 5 or 10. Discuss how to recognise multiples of these numbers (multiples of 2 are even, multiples of 5 end in 5 or 0, and multiples of 10 end in 0).
- Demonstrate the following two-player game: Players take it in turns to throw a six-sided dice marked 2, 2, 5, 5, 10, 10. The player who threw the dice puts a counter on a hundred square covering a multiple of the number they threw. The winner is the first player to get four counters in a row.
- Organise the learners into pairs to play the game, giving each pair a small hundred square and plenty of counters in two colours.

Plenary

- Ask: *In the game are there any numbers that are better to throw than others?* (Yes, 2.) *Why?* (Because this gives you the most numbers to choose from when placing your counter.)
- Discuss game strategies (such as trying to block your opponent) and whether going first affects your chance of winning.

Success criteria

Ask the learners:

- Throw two dice: one ten-sided dice marked 0 to 9 and one of the dice used in the game (a six-sided dice marked 2, 2, 5, 5, 10, 10). Multiply the two numbers together. What is the answer?
- What three other facts follow from 3 × 7 = 21?

Ideas for differentiation

Support: In the game, give these learners half a hundred square (1 to 50) instead of a whole one.

Extension: Give these learners a limited time for each move in the game (for example by using a timer).

Tables chart

2 times table	3 times table	5 times table	10 times table
2 × 1 = 2	3 × 1 = 3	5 × 1 = 5	10 × 1 = 10
2 × 2 = 4	3 × 2 = 6	5 × 2 = 10	10 × 2 = 20
2 × 3 = 6	3 × 3 = 9	5 × 3 = 15	10 × 3 = 30
2 × 4 = 8	3 × 4 = 12	5 × 4 = 20	10 × 4 = 40
2 × 5 = 10	3 × 5 = 15	5 × 5 = 25	10 × 5 = 50
2 × 6 = 12	3 × 6 = 18	5 × 6 = 30	10 × 6 = 60
2 × 7 = 14	3 × 7 = 21	5 × 7 = 35	10 × 7 = 70
2 × 8 = 16	3 × 8 = 24	5 × 8 = 40	10 × 8 = 80
2 × 9 = 18	3 × 9 = 27	5 × 9 = 45	10 × 9 = 90
2 × 10 = 20	3 × 10 = 30	5 × 10 = 50	10 × 10 = 100

Cambridge Primary: Ready to Go Lessons for Maths Stage 3 © Hodder & Stoughton Ltd 2013

Multiplication and division 2

Learning objectives

- Begin to know the 4× table. (3Nc4)
- Make sense of and solve word problems and begin to represent them. (3Pt3)
- Consider whether an answer is reasonable. (3Pt12)
- Make up a number story to go with a calculation. (3Ps1)
- Explore and solve number problems and puzzles. (3Ps3)

Resources

Large and small number lines (0 to 40, every number labelled); counters; photocopiable page 30.

Starter

- Display a large 0 to 40 number line. Demonstrate counting on in fours on the number line to generate the first few numbers in the 4 times table. Indicate each jump with an arc and circle each multiple of 4. Ask a learner to complete the pattern to 40.
- Chant the 4 times table together, referring to the number line.
- Call out multiplications from the 4 times table in a random order. Ask the learners to call out the product (answer), using the number line for support. Extend to include divisions (for example: *How many fours are there in 28?*)

Main activities

- Display a copy of photocopiable page 30. Work through a couple of the problems together, asking: *What calculation do you need to do to solve this problem?* Do the calculation. Ask: *What's the answer? How did you work it out?* Discuss materials that may help the calculation (for example counting on a number line, using counters or drawing a diagram). Ask: *Is your answer reasonable? How do you know?*

- Explain to the learners what they need to do for question 9 (write their own problems based on given calculations).
- Organise the learners into pairs and give each pair photocopiable page 30. Ensure every table has some 0 to 40 number lines and some counters, in case the learners want to use them to support their calculations.

Plenary

- Choose one of the problems not worked through in the main activity, and work through it in a similar way.
- Ask volunteers to share the problems they have written, and ask the rest of the class to solve them. Discuss their answers.

Success criteria

Ask the learners:

- What calculation did you need to do in order to solve this problem?
- Explain how you worked out the answer.
- Is your answer reasonable? How do you know?
- Make up a number problem for the calculation $20 \div 4$.

Ideas for differentiation

Support: For these learners, model how to use equipment to support multiplication and division calculations. You may prefer to demonstrate a single technique, or to demonstrate several and allow the learners to choose.

Extension: Challenge these learners to devise some more problems that involve multiplying and dividing by 6, 7, 8 or 9.

Name: _____

Supermarket problems

1. In the supermarket there are six groups of three shoppers.
    How many shoppers are in the supermarket altogether?

    _____

2. Eight bags of rice weigh a total of 16 kg. How much does each bag of rice weigh?

    _____

3. A bunch of flowers costs $5. Sam bought seven bunches of flowers.
    How much money did he spend?

    _____

4. A box of cereal is 10 cm wide. How many boxes will fit on a shelf
    that is 80 cm long?

    _____

5. A carton of milk holds 2 litres. How much milk do seven cartons hold?

    _____

6. Mira divides 15 cans equally among five shelves.
    How many cans does she put on each shelf?

    _____

7. Water filters cost $10. How much would you pay for three water filters?

    _____

8. The supermarket has eight aisles, with four shelves in each aisle.
    How many shelves are there altogether?

    _____

9. Write a problem for each of these calculations.

    a) 9 × 3 _____

        _____

    b) 8 × 5 _____

        _____

    c) 36 ÷ 4 _____

        _____

Cambridge Primary: Ready to Go Lessons for Maths Stage 3 © Hodder & Stoughton Ltd 2013

Doubling and halving 1

Learning objectives

- Understand the relationship between doubling and halving. (3Nc19)
- Choose appropriate mental strategies to carry out calculations. (3Pt1)
- Explain a choice of calculation strategy and show how the answer was worked out. (3Ps2)
- Identify simple relationships between numbers. (3Ps6)

Resources

Large cards made from photocopiable page 32; price-labelled items (enough for about two per learner) – prices between $2 and $98 (even numbers only); blank labels (tie-on or stick-on).

Starter

- Prepare the doubling and halving cards from photocopiable page 32, enlarging onto A3 card (or bigger) and cutting out.
- Hold up the cards one at a time, asking the learners to call out the answer. Make sure the first half-dozen cards feature only numbers below 20. Through questioning, make explicit the relationship between doubling and halving. When you reach the first card featuring larger numbers, ask: *How did you work out the answer?* Make explicit the relationship between doubling multiples of 10 and doubling single-digit numbers. (When doubling multiples of 10 the answer is 10 times bigger than for the corresponding single-digit number, for example double 40 = 80, compared to double 4 = 8.)

Main activities

- Focus on mental methods, but model using jottings for support, and encourage the learners to do the same.
- Display a two-digit number less than 50 that is not a multiple of 10 (for example 38). Ask the learners to double the number and explain their method. Repeat for similar numbers.

- Demonstrate how to halve multiples of 10 in which the tens digit is odd (for example 50 = 40 + 10; half of 40 is 20 and half of 10 is 5, so half of 50 is 20 + 5, which is 25).
- Display a two-digit number in which both digits are even (for example 86). Ask the learners to halve the number and explain their method. Repeat for a number in which only the units digit is even (for example 78).
- Distribute price-labelled items to each table. Explain that the items are going to be in a half-price sale, and ask the learners to write a new price label for each item. Swap re-priced items between tables and ask the learners to work out the original prices.

Plenary

- Use the plenary to extend to doubling larger two-digit numbers and halving numbers to 200.
- Ask: *What's double 93? How did you work it out?*
- Ask: *What's half of 170? How did you work it out?*

Success criteria

Ask the learners:

- Which doubling fact can help you work out the answer to double 70?
- What's double 26? Explain how you worked out the answer.
- What's half of 72? Explain how you worked out the answer.

Ideas for differentiation

Support: In the final main activity, sit these learners together and give them items with prices less than $30. Sit with them, prompting and supporting their calculations.

Extension: In the final main activity, sit these learners together and give them items with prices more than $50.

Doubling and halving cards 1

Double 1	Half of 2	Double 2	Half of 4
Double 3	Half of 6	Double 4	Half of 8
Double 5	Half of 10	Double 6	Half of 12
Double 7	Half of 14	Double 8	Half of 16
Double 9	Half of 18	Double 10	Half of 20
Double 20	Half of 40	Double 30	Half of 60
Double 40	Half of 80	Double 50	Half of 100
Double 60	Half of 120	Double 70	Half of 140
Double 80	Half of 160	Double 90	Half of 180
Double 100	Half of 200		

Unit assessment

- Which place value cards do you need to make 385?
- How do you multiply a number by 10?
- Can you write a list, draw a table or draw a diagram to show pairs of multiples of 100 that total 1000 (for example 100 + 900 = 1000)?

- Can you add together 6, 9, and 4? Explain how you worked out the answer.
- Can you write a problem for this calculation: 24 ÷ 3?
- What's half of 120? How did you work it out?

Summative assessment activities

Observe the learners while they take part in these games. You will quickly be able to identify those who appear to be confident and those who may need additional support.

Number line game

This game assesses the learners' knowledge of place value to 1000.

You will need:

Ten-sided dice (0 to 9); 0 to 1000 number lines labelled every 100, with non-labelled divisions every 10; coloured pencils.

What to do

- Organise the learners into groups of three to five.
- Give each group three ten-sided dice and a number line, and give each player a different-coloured pencil. Tell the players to take it in turns to roll three dice, make the largest possible number and the smallest possible number, and mark these on the number line in coloured pencil, saying each number out loud.
- The game ends when every player has had five turns. The largest number scores 3 points, then 2 and 1 for the next largest numbers. The smallest numbers are scored in the same way. The winner is the player with the most points.

Addition facts to 20 game

This game assesses the learners' recall of addition facts up to 20.

You will need:

Sticky notes in two colours.

What to do

- Divide the class into two teams and sit each team in a line, one behind another. At the front of each line place a pencil and a stack of sticky notes in a particular colour.
- Call out an addition with a total up to 20. The learner at the front of the line writes their name and the total on a sticky note, and races to give it to you. Put the first correct answer in one pile. Put other correct answers in a second pile and incorrect answers in a third pile.
- When every learner has had a go, count the number of sticky notes of each colour in the first pile. The team with the most sticky notes wins.

Distribute photocopiable page 34. Ask the learners to read the questions and write the answers. They should work independently.

Name: _____

Number problems

1. Write four hundred and six in figures.

    _____

2. What is the value of the 5 in 259?

    _____

3. What number is 10 less than 341?

    _____

4. What number is 100 more than 715?

    _____

5. Nadia made the number 580 by multiplying a starting number by 10. What was the starting number?

    _____

6. Write the next three numbers in this number pattern:

    28, 24, 20, 16, _____ , _____ , _____

7. Is 978 a multiple of 2, 5 or 10?

    _____

8. Write the missing number. ☐ + 85 = 100

9. Subtract 30 from 273.

10. A tennis racket normally costs $54.

    What price will it be in a half-price sale?

    _____

Unit 1B: Geometry and problem solving

Co-ordinates 1

Learning objectives

- Use the language of position, direction and movement, including clockwise and anti-clockwise. (3Gp1)

Resources

Photocopiable page 36; pencils; plain paper.

Starter

- Display a copy of photocopiable page 36.
- Give the co-ordinates of a square on the map and ask the learners what can be found there, for example: *What is in square 7?* Repeat for a few other features.
- Name one of the features on the map, asking the learners to give its co-ordinates, for example: *Which square is Skull Island in?* Repeat for a few other features.
- Ask the learners to give directions from one feature to another, for example: *Describe how to get from Skeleton Cliff to the Dark Forest.* (Go two squares east and one square south.) Repeat for a few other pairs of features.

Main activities

- Ask the learners to create their own treasure maps that use a grid with letters and numbers. You may want to ask some learners to work independently and some to work in pairs.
- Ask each learner or pair to write six questions about their map's features, similar to the questions asked in the starter activity. When they have finished writing their questions, ask them to give the questions (together with their map) to another learner or pair to answer.

Plenary

- Ask the learners questions about clockwise and anti-clockwise turns in units of a quarter turn, for example: *You stand on Lookout Point and face east. You turn anti-clockwise through a quarter of a turn. Which direction are you facing now?* (North.) *If I face south and then turn through three quarters of a turn in a clockwise direction, which direction will I be facing?* (East.)

Success criteria

Ask the learners:

- (Pointing to an empty square on the treasure map:) Which square is this?
- Can you point to F6?
- (Pointing to two locations on the treasure map:) Describe how to get from here to here.

Ideas for differentiation

Support: Ask these learners to write fewer (for example four) questions in the main activity.

Extension: Ask these learners to write more (for example eight) questions in the main activity.

Treasure map

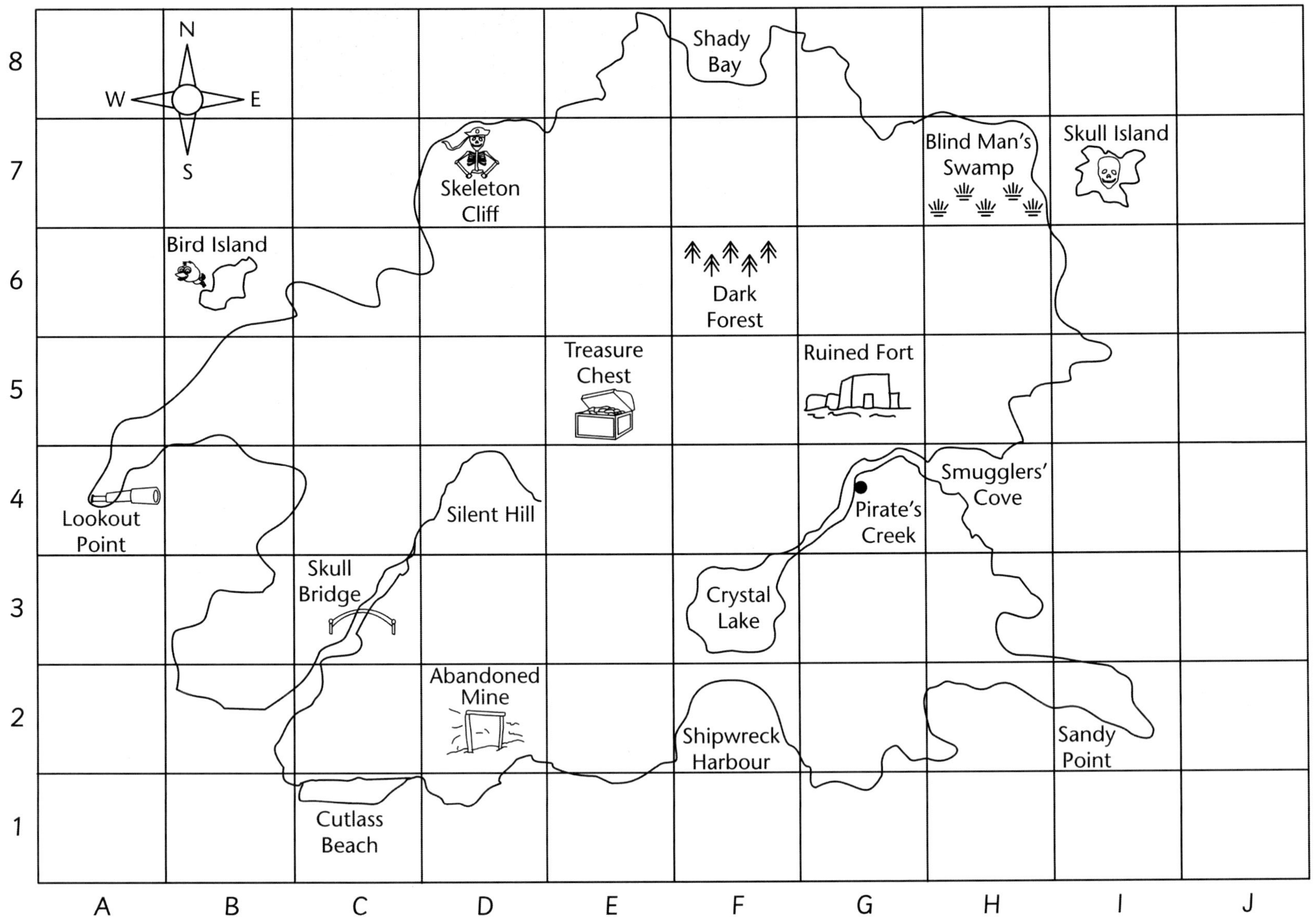

Compass rose: N, E, S, W

Bird Island

Lookout Point

Skeleton Cliff

Shady Bay

Treasure Chest

Dark Forest

Silent Hill

Skull Bridge

Ruined Fort

Blind Man's Swamp

Skull Island

Crystal Lake

Smugglers' Cove

Pirate's Creek

Abandoned Mine

Cutlass Beach

Shipwreck Harbour

Sandy Point

A B C D E F G H I J

8 7 6 5 4 3 2 1

Cambridge Primary: Ready to Go Lessons for Maths Stage 3 © Hodder & Stoughton Ltd 2013

Symmetry 1

- Draw and complete 2D shapes with reflective symmetry and draw reflections of shapes. (3Gs5)
- Identify simple relationships between shapes, e.g. these shapes all have the same number of lines of symmetry. (3Ps7)

Photocopiable page 38; one large set of 2D shapes cut out of coloured paper (for example rectangle, square, kite, rhombus, parallelogram, equilateral triangle, isosceles triangle, scalene triangle, regular pentagon, regular hexagon); 2D shapes; mirrors; examples of symmetrical patterns from the internet; plain paper; coloured pens or pencils.

Starter

- Display a copy of photocopiable page 38 and give each pair of learners a copy. Explain that the unbroken lines in each diagram show half a shape. The dotted line is a mirror line or line of symmetry. Ask the learners to finish drawing each shape and write its name (if they know it).
- Complete the display copy together.

Main activities

- Display the coloured paper shapes. Ask: *What do we mean when we say a shape has symmetry?* Discuss the learners' suggestions.
- Point to one of the shapes, asking: *How many lines of symmetry does this shape have? How could we find out?* Discuss the learners' ideas, and try them out on a couple of the shapes. Ensure you cover both folding the shape in half and using a mirror.

- Organise the learners into groups of three or four and give each group a set of shapes. Ask them to sort the shapes into groups according to their number of lines of symmetry and record their groupings.
- Display some examples of patterns with reflective symmetry. Discuss the use of both shape and colour, and identify the number of lines of symmetry in each pattern. Challenge the learners to combine various shapes to make their own symmetrical patterns.

Plenary

- Discuss the learners' groupings of shapes according to lines of symmetry.
- Ask the learners to share and discuss the symmetrical patterns they've made.

Ask the learners:

- Can you draw a shape that is symmetrical?
- How many lines of symmetry does the shape have?
- Tell me about the symmetrical pattern you've drawn. How many lines of symmetry does it have?

Support: Group these learners together and work with them to help them sort shapes into symmetry groups.

Extension: In the sorting activity, group these learners together and give them a wider range of shapes to sort.

Name: _____

Shape reflections

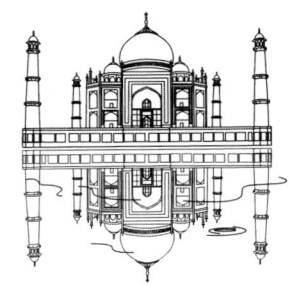

1. Complete each shape along the dotted line of symmetry.
2. Write the name of each shape.

a)

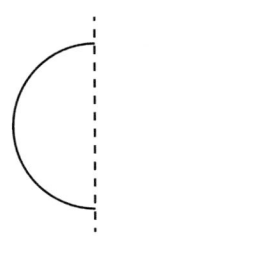

_____

b)

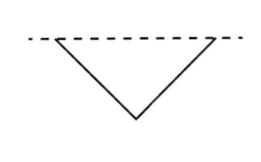

_____

c)

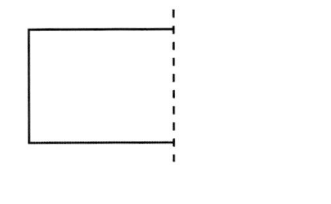

_____

d)

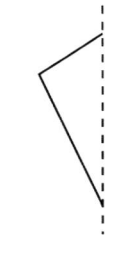

_____

e)

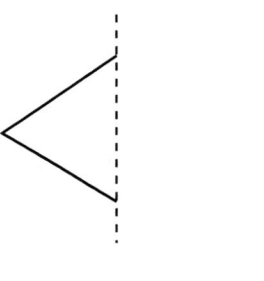

_____

f)

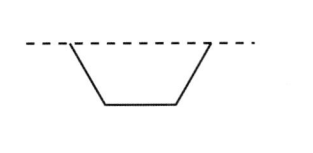

_____

g)

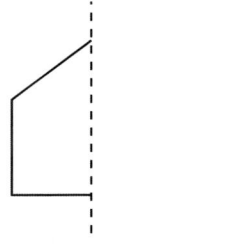

_____

h)

_____

Cambridge Primary: Ready to Go Lessons for Maths Stage 3 © Hodder & Stoughton Ltd 2013

Angles 1

Learning objectives

- Classify 2D shapes according to the number of sides, vertices and right angles. (3Gs2)
- Identify right angles in 2D shapes. (3Gs8)

Resources

A large clock face with moveable hands; photocopiable page 40; 2D shapes (large and small); large right-angle measurer (a large circle cut out of coloured paper and folded into quarters); plain paper; compasses; scissors.

Starter

- Explain that an angle is a measure of the amount of turn. Demonstrate a clock hand turning through an acute angle. Demonstrate how to draw the angle the clock hand has made using two straight lines and a curved arrow to show the movement.
- Move the clock hand through various angles. Ask the learners to sketch each angle. When you first move the clock hand through a quarter of a circle, explain that this is called a right angle and that when drawn, the inside is marked with a square corner.

Main activities

- Draw a right angle on the board, marking it with a square corner and labelling it 'right angle'.
- Show the learners objects in the classroom that have right angles in them, and point to the right angles. Ask the learners to identify more right angles around the classroom.
- Display an angle cut out from photocopiable page 40. Ask whether it is less than, equal to or greater than a right angle. Confirm the size by using the large right-angle measurer. Repeat for several angles on photocopiable page 40.
- Demonstrate how to make a right-angle measurer by drawing a circle with a pair of compasses, cutting it out, then folding it into quarters.
- Ask every learner to make a right-angle measurer and then use it to measure angles in 2D shapes. Ask the learners to record their findings.

Plenary

- Draw this table on the board:

No right angles	1 right angle	2 right angles	3 right angles	4 right angles

- Hold up large 2D shapes one at a time, asking the learners to place each shape in the correct column in the table.
- Ask: *How many of the shapes have 2 or 3 right angles?*
- Challenge the learners to draw a different shape that has 2 or 3 right angles.

Success criteria

Ask the learners:

- Is this angle less than, more than or equal to a right angle?
- How many right angles does this shape have?
- Name a shape with 4 right angles.
- Name a shape with no right angles.

Ideas for differentiation

Support: Assist these learners when they are making their angle measurers to ensure they produce a true right angle.

Extension: Ask these learners to group 2D shapes according to the number of acute or obtuse angles they have.

Angles

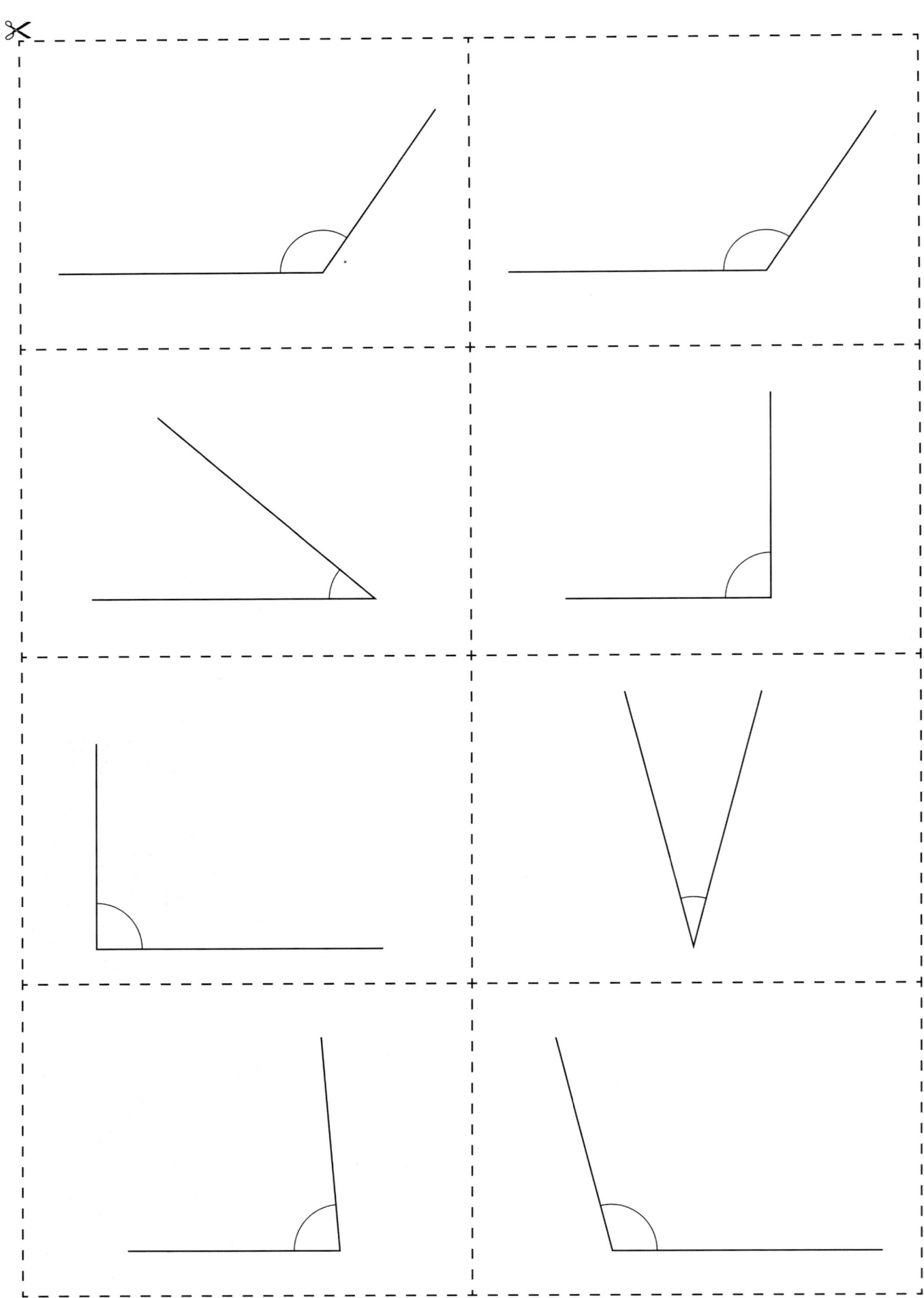

Cambridge Primary: Ready to Go Lessons for Maths Stage 3 © Hodder & Stoughton Ltd 2013

2D shapes 1

Learning objectives

- Identify, describe and draw regular and irregular 2D shapes including pentagons, hexagons, octagons and semi-circles. (3Gs1)
- Classify 2D shapes according to the number of sides, vertices and right angles. (3Gs2)
- Recognise the relationships between different 2D shapes. (3Pt8)

Resources

Photocopiable page 44; pencils; plain paper; photocopiable page 42; mirrors.

Starter

- Display a set of 2D shape cards made by enlarging photocopiable page 44 onto A3 card and cutting out.
- Give each learner paper and a pencil.
- Describe the properties of one of the shapes. Ask the learners to sketch the shape, and name the shape they have drawn. Point to the shape in the display and label it with its name. Repeat for other shapes.
- Ask the learners to continue the activity in pairs, one choosing and describing a shape and the other drawing and naming it, and then swapping roles.

Main activities

- Ask the learners to sort 2D shape cards made from photocopiable page 44 according to their number of a) sides, b) vertices, c) right angles and d) lines of symmetry.
- Ask: *Are there any shapes that don't have any vertices? What is the shape with the greatest number of lines of symmetry? How many does it have?*

- Play this game for four players: Shuffle together two sets of the 2D shape cards and deal them all out. Shuffle a set of property cards made from photocopiable page 42 and place them in a face-down pile. Ask the players to take it in turns to turn over the top property card. If they have a card in their hand that matches the property on the card, they should play their card and name the shape. Otherwise, play passes to the left. The winner is the first player to have no cards left.

Plenary

- Draw the following Venn diagram on the board. Remind the learners how this type of diagram works.

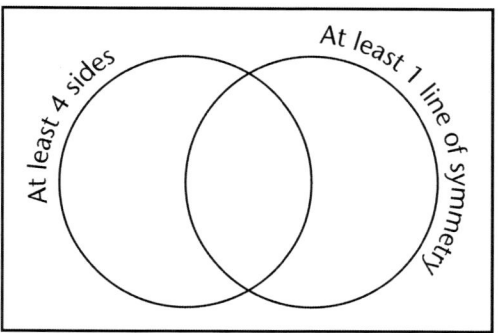

- Ask a volunteer to place a shape in the correct position on the diagram. Repeat several times with different shapes and different volunteers.

Success criteria

Ask the learners:

- Draw an irregular hexagon.
- Name this shape. Describe one of its properties.
- Sort these shapes according to their number of vertices.

Ideas for differentiation

Support: Provide mirrors during the main activities to help these learners visualise the lines of symmetry.

Extension: In the first main activity, challenge these learners to sort the shapes in a fifth way according to their own ideas.

Property cards

3 sides	4 sides	5 sides
6 sides	8 sides	no right angles
1 right angle	more than 1 right angle	no equal sides
1 pair of equal sides	2 pairs of equal sides	all sides equal
3 vertices	4 vertices	5 vertices
6 vertices	8 vertices	no vertices
no lines of symmetry	1 line of symmetry	more than 1 line of symmetry

2D shapes 2

- Identify, describe and draw regular and irregular 2D shapes including pentagons, hexagons, octagons and semi-circles. (3Gs1)
- Use a set square to draw right angles. (3Gp3)
- Recognise the relationships between different 2D shapes. (3Pt8)

Photocopiable page 44; pencils; plain paper; rulers; set squares; compasses; erasers; coloured paper (A3 or larger); scissors; felt-tip pens; materials for creating a wall display.

Starter

- Shuffle a set of cards made by enlarging photocopiable page 44 onto A3 card and cutting out. Hold the cards up one at a time, asking the learners to call out the name of each shape. Demand specific rather than general names (for example 'isosceles triangle' rather than 'triangle').
- Give each learner paper and a pencil. Reshuffle the cards. Look at a card without displaying it. Name the shape on the card. Ask the learners to draw a sketch of the shape. Confirm the correct answer by showing the card. For each shape, ask the learners to describe one or more of its properties.

Main activities

- Give each learner a pair of compasses, a ruler, a set square and several sheets of plain paper.
- Demonstrate drawing each of the following in turn, and then ask the learners to practise drawing them:
  - a circle and a semi-circle using a pair of compasses and a ruler
  - a square and a rectangle using the set square
  - a kite using the ruler.

- Ask the learners to draw irregular pentagons, hexagons and octagons using a ruler.
- Give out the large coloured paper and the scissors. Ask the learners to choose two of the shapes they have drawn, draw large versions of the shapes on coloured paper, cut them out and label each shape with its name.

Plenary

- Hold up two coloured shapes. Ask the learners to name the shapes and then ask: *What is the same about these two shapes? What is different about them?* Repeat for different pairs of shapes.
- Begin making the shapes into a wall display.

Ask the learners:

- Can you draw a square / a semi-circle / a kite?
- (Pointing to two shapes:) What is the same about these shapes? What is different about them?

Support: Seat these learners together and work with them when they are practising drawing shapes.

Extension: Challenge these learners to draw a regular hexagon. If they need help getting started, give them equilateral triangle shape pieces, and suggest they start by putting them together to form a hexagon.

2D shape cards

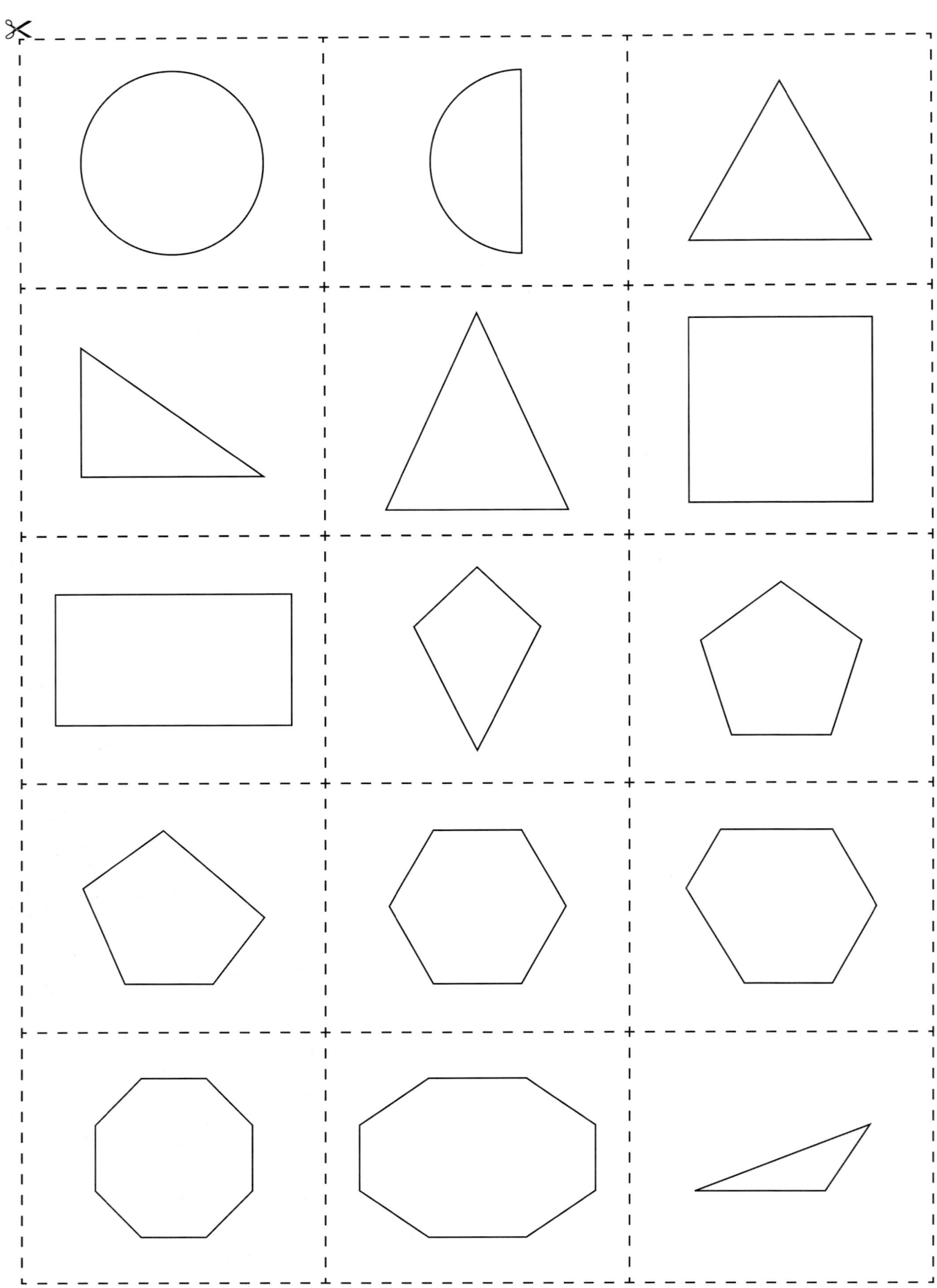

Cambridge Primary: Ready to Go Lessons for Maths Stage 3 © Hodder & Stoughton Ltd 2013

3D shapes 1

- Identify, describe and make 3D shapes including pyramids and prisms; investigate which nets will make a cube. (3Gs3)
- Classify 3D shapes according to the number and shape of faces, number of vertices and edges. (3Gs4)
- Relate 2D shapes and 3D solids to drawings of them. (3Gs6)

Photocopiable page 46; 3D shapes: set A (cube, cuboid, sphere, cylinder, cone) and set B (triangular prism, pentagonal prism, hexagonal prism, square-based pyramid, triangular-based pyramid, pentagonal-based pyramid, hexagonal-based pyramid); coloured sticky dots.

Starter

- Remind the learners of the terms 'face' (a flat surface), 'edge' and 'vertex' (the point where three or more faces meet). Point to an example of each feature on a 3D shape.
- Shuffle a set of cards made by enlarging photocopiable page 46 onto A3 card and cutting out, and hold them up one at a time. Ask the learners to say the appropriate word and do the appropriate action to go with it, for example FACE – place the palms of the hands together; EDGE – place the little finger edge of one hand on top of the index finger edge of the other; VERTEX – bring the fingertips together to form a pointed inverted 'v'. Keep the pace brisk.

Main activities

- Organise the learners into groups of four. Give each group the set A shapes. Hold up the set A shapes one at a time, asking the learners to name them. Ask them to count the number of faces, edges and vertices in each shape and record the information in a table.
- Give each group the set B shapes.
- Show and name the triangular prism from set B and define a prism. It has one pair of identical faces that are parallel to each other (the 'ends').

The two ends are joined by another set of faces. These faces are usually rectangles, but may also be squares or parallelograms. The shape of the prism's ends gives the prism its name. A prism with ends that are triangles is a triangular prism. Ask the learners to identify other shapes that are prisms, and suggest what each might be called.

- Show and name the square-based pyramid from set B and define a pyramid. It has a set of triangular faces that meet together at a single vertex. All the triangular faces connect to a single face called the base. The base is a polygon (a shape with straight sides). The shape of the pyramid's base gives the pyramid its name. So a pyramid with a base that is a square is called a square-based pyramid. Ask the learners to identify other shapes that are pyramids.
- Ask the learners to sort the shapes into groups of their own devising. Discuss various properties the shapes could be sorted by (for example number of faces, vertices, edges; shape of faces).

Plenary

- Display the cards from the starter activity. Hold up a 3D shape from set A or B (not the sphere or cone). Ask a volunteer to name the shape and find a card that shows it.

Ask the learners:

- Show me the cuboid. How many faces does it have? Tell me something else about it.
- Show me the triangular prism. How many vertices does it have? Tell me something else about it.

Support: In the first main activity give these learners sticky dots in three colours to mark counted faces, edges and vertices.

Extension: Challenge these learners to record the number of faces, edges and vertices in various pyramids, record their findings in a table and describe any patterns they see in the numbers.

Face, edge or vertex?

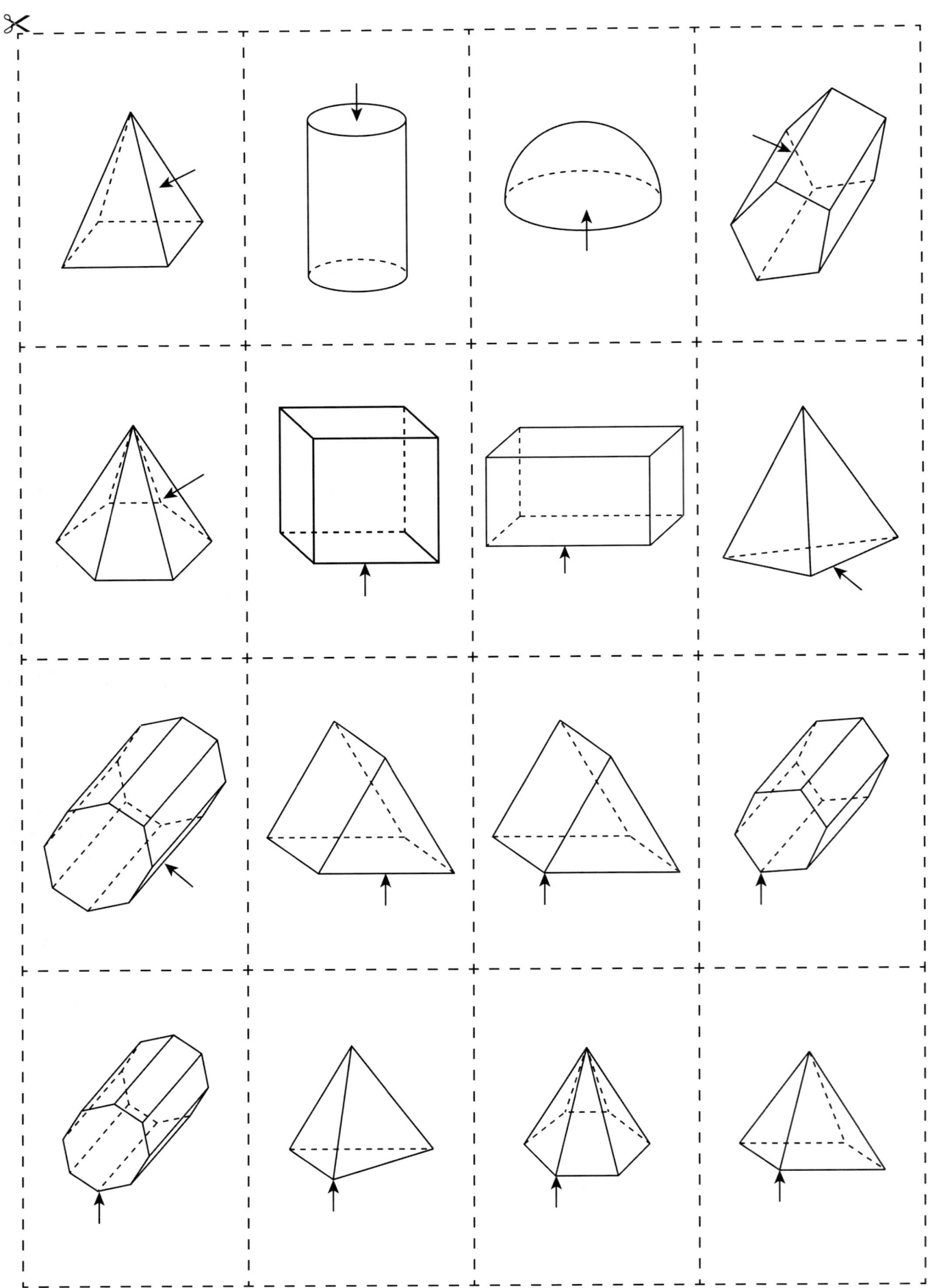

3D shapes 2

Learning objectives

- Identify, describe and make 3D shapes including pyramids and prisms; investigate which nets will make a cube. (3Gs3)
- Identify 2D and 3D shapes, lines of symmetry and right angles in the environment. (3Gs7)
- Identify the differences and similarities between different 3D shapes. (3Pt9)

Resources

3D shapes (for example a sphere, a cone, a cylinder, a variety of pyramids and a variety of prisms, including a cube and a cuboid); pyramids and prisms made from art straws and modelling clay; art straws; modelling clay; scissors; rulers; large net of a cube made from card; photocopiable page 48; squared paper.

Starter

- Hold up 3D shapes two at a time, asking the learners to name both shapes, and describe as many things as they can that are the same about the two shapes. Then ask the learners to describe as many things as they can that are different about the two shapes. Repeat for three or four different pairs of shapes.
- Hold up the art straw pyramids and prisms one at a time, asking the learners to name each shape and match it to the solid version of the shape.
- Ask the learners to identify objects (or parts of objects) in the classroom that are examples of named 3D shapes.

Main activities

- Organise the learners into mixed-ability groups. Give each group a selection of 3D shapes (including pyramids and prisms), art straws, modelling clay, scissors and rulers. Ask the learners to use the art straws and modelling clay to make skeletons of the shapes.

- Revise the concept of nets, and demonstrate how a card net folds up to make a cube. Display a copy of photocopiable page 48. Ask the learners to visualise cutting out each shape and folding it. Which of them will make a cube? Ask the learners to record their predictions.
- Ask the learners to draw each shape on squared paper, cut it out and then try folding it into a cube.

Plenary

- Discuss which shapes make a cube. (All of them apart from A, C, E and I.) Ask: *Were your predictions correct?*

Success criteria

Ask the learners:

- What can you see in the classroom that is the same shape as this?
- How many different nets have you found that will make a cube?
- Choose two 3D shapes. What is the same about them? What is different?

Ideas for differentiation

Support: Give these learners their own copy of photocopiable page 48 so that they can simply cut out the existing shapes instead of drawing their own.

Extension: Challenge these learners to make a triangular prism or a square-based pyramid out of card by first designing a net on squared paper.

Cube nets

Which of these shapes can be folded to make a cube? Copy each one onto squared paper to find your answer.

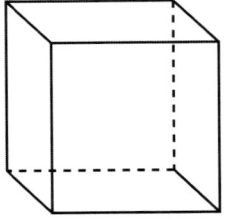

a)

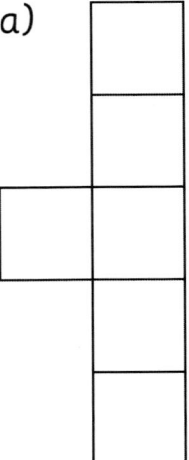

b)

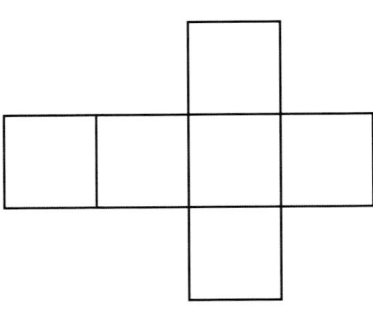

c)

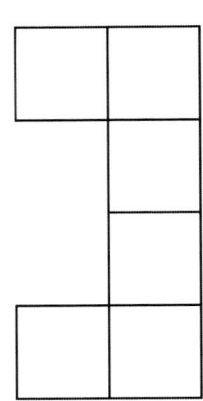

d)

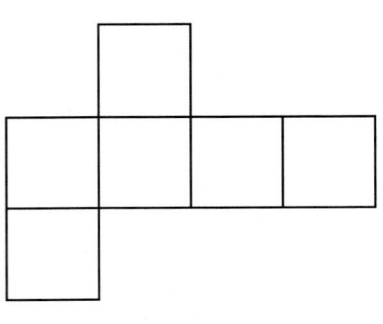

e)

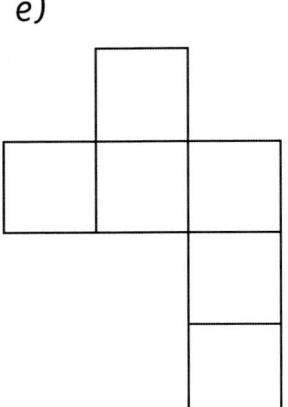

f)

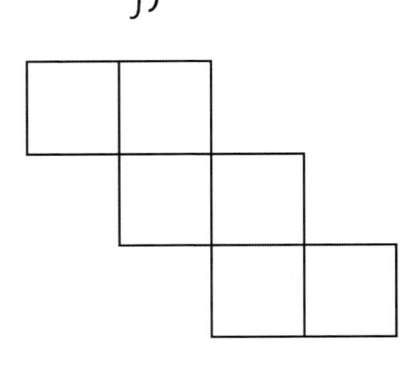

g)

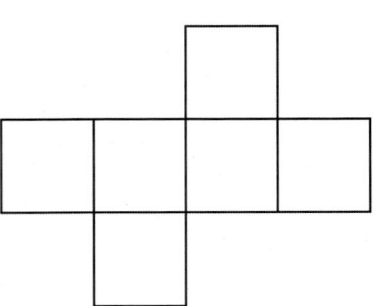

h)

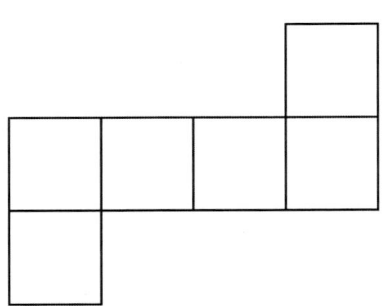

i)

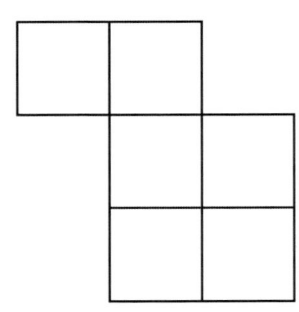

Unit assessment

- Can you make an accurate drawing of a symmetrical 2D shape and label it with its name?
- How many lines of symmetry does the shape have? Draw them in.

- Can you describe the other properties of the shape you have drawn?
- Can you make a skeleton model of a 3D shape and label it with its name?
- Can you describe the properties of the 3D shape you made?

Summative assessment activities

Observe the learners while they take part in these activities. You will quickly be able to identify those who appear to be confident and those who may need additional support.

Angles and symmetry

This activity assesses the learners' ability to identify right angles and lines of symmetry in 2D shapes.

You will need:

Large cards with regular and irregular 2D shapes drawn on them (including shapes with between 0 and 4 right angles and shapes with between 0 and 8 lines of symmetry); 0 to 9 number fans.

What to do

- Give each learner a 0 to 9 number fan.
- Hold up a card, asking: *How many right angles are there in this shape?* The learners must show the answer on their number fan. Confirm the right answer, pointing to all the right angles in the shape.
- Still holding up the same shape, ask: *How many lines of symmetry does this shape have?* Again, the learners show the answer on their number fan. Confirm the right answer, 'drawing in' with your finger all the lines of symmetry in the shape.
- Repeat for the rest of the cards.

Shape walk

This activity assesses the learners' ability to identify 2D and 3D shapes, lines of symmetry and right angles in the environment.

You will need:

Sketchbooks, pencils, notebooks, cameras (optional).

What to do

- Go on a shape-focused Maths walk around the school or the locality. Ask the learners to look for 2D and 3D shapes, lines of symmetry and right angles. The learners record their observations by drawing sketches, making notes and taking photographs.
- Discuss the information collected on the Maths walk. Ask the learners to decide how to present their observations (for example in a book, in a wall display, at an assembly or in a film presentation).

Give each learner an enlarged, A3 copy of photocopiable page 50 and a selection of around a dozen 3D shapes. Ask the learners to place each shape in the correct region of the diagram. When the learners have finished sorting the shapes, ask them to record their groupings by writing the name of each shape in the appropriate region of the diagram.

Ask those who have finished to draw another Venn diagram on the back of photocopiable page 50, labelling each circle with their own choice of property. Ask them to re-sort the shapes according to these new criteria, and record the new groupings.

  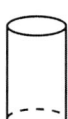 

Sorting 3D shapes

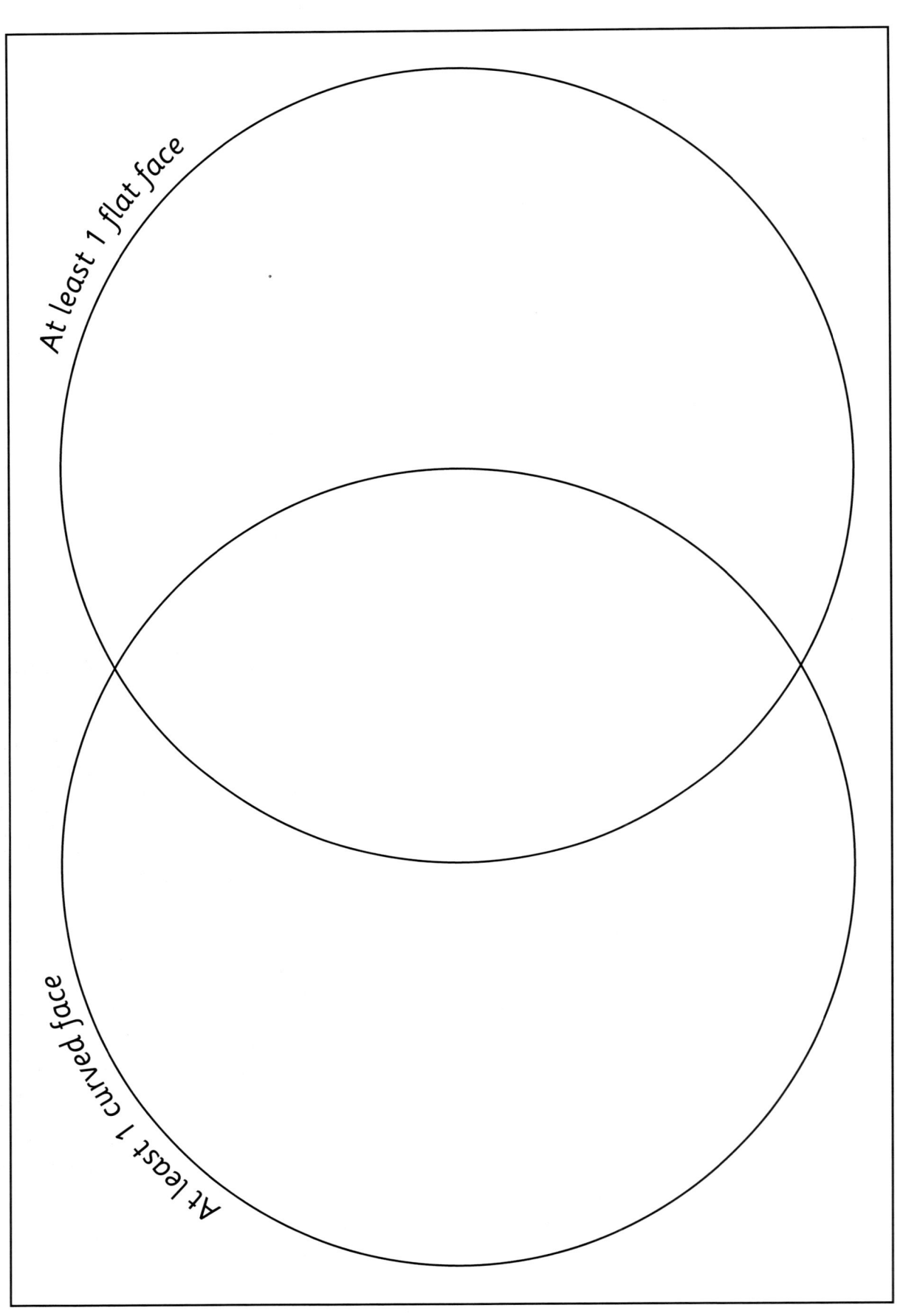

At least 1 flat face

At least 1 curved face

Money 1

Learning objectives

● Consolidate using money notation. (3Mm1)
● Use addition and subtraction facts with a total of 100 to find change. (3Mm2)

Resources

Several sets of number cards showing the multiples of 5, from 5 to 95 (two cards per set showing 50); US currency (1c, 5c, 10c, 25c and 50c coins, and $1 bills – real if possible); one large and plenty of small 0 to 100 number lines; photocopiable page 52.

Starter

- Organise the learners into mixed-ability teams. Give each team a set of 'multiples of 5' number cards. Teams must race against each other to match up pairs of cards that total 100. The first team to match all their cards correctly scores 1 point.
- Repeat several times, each team shuffling their cards thoroughly and passing them to another team (this ensures the cards get shuffled properly!).

Main activities

- Display coins one at a time, asking the learners to give their names and values (a 1c coin is a penny or a cent, a 5c coin is a nickel, a 10c coin is a dime, a 25c coin is a quarter and a 50c coin is a 50-cent piece or a half-dollar.)
- Display a $1 bill. Ask: *How many cents make one dollar?* (100.) Holding up a coin, ask: *How many of these coins make one dollar? How did you work it out?* Repeat for all the coins.

- Display a copy of photocopiable page 52 and the large 0 to 100 number line. Demonstrate finding change from $1 by counting up on a number line (for example for a spend of 77 cents: *From 77 to 80 is 3, then from 80 to 100 is 20, so I need 23 cents change*). Then demonstrate counting out the correct change in coins. (For example, counting out two 10c coins and three 1c coins, say: *Ten, twenty, twenty-one, twenty-two, twenty-three.* Ask the learners to name the coins you have used.) Organise the learners into pairs and give each pair coins, a small number line and their own copy of photocopiable page 52 and ask them to work through the questions.

Plenary

- Take the learners' answers to the questions on photocopiable page 52. Stress the fact that there are often many different ways to make the same value of change.

Success criteria

Ask the learners:

● Name this coin. What is it worth? How many of them make $1?
● What change would you get from $1 if you bought an eraser for 67c? Make the change in coins.

Ideas for differentiation

Support: Group these learners together and work through several more examples with them, leading them through the process step by step. Allow them to attempt the last few questions independently.

Extension: Once these learners are confident using the number line, challenge them to do without it, and to find change by counting up mentally.

Name: _____

Finding change from $1

Look at each item below and work out how much change you would get from $1.

Write:

a) the number of cents change

b) one way of making that amount of change in coins.

The first question has been answered for you.

1.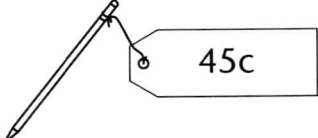

   a) 55c change

   b) 25c + 25c + 5c

2.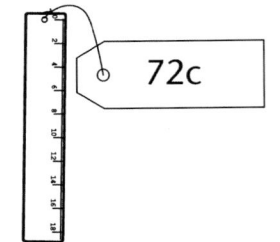

   a) _____

   b) _____

3.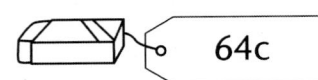

   a) _____

   b) _____

4.

   a) _____

   b) _____

5.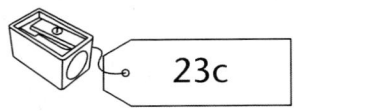

   a) _____

   b) _____

6.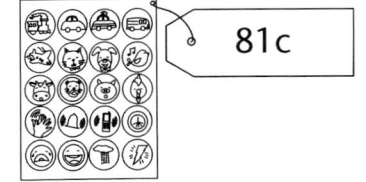

   a) _____

   b) _____

7.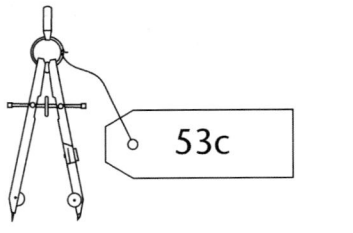

   a) _____

   b) _____

8.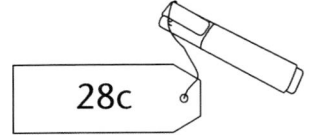

   a) _____

   b) _____

Cambridge Primary: Ready to Go Lessons for Maths Stage 3 © Hodder & Stoughton Ltd 2013

Money 2

Learning objectives

- Choose appropriate mental strategies to carry out calculations. (3Pt1)
- Make up a number story to go with a calculation, including in the context of money. (3Ps1)

Resources

Pencils; plain paper; photocopiable page 54.

Starter

- Display an addition sum with a total of less than 100 (for example 56 + 27). Ask the learners to do the calculation using whichever method they prefer, and write down the answer. Take in answers and discuss strategies used (for example sketching a number line, or partitioning and recombining). Repeat for several additions.
- Display a subtraction containing two two-digit numbers (for example 62 − 34). Discuss possible calculation strategies (essentially the same strategies as used for addition). Give the learners several subtractions to work out.

Main activities

- Display a copy of photocopiable page 54. Ask the learners to suggest a price for each type of sweet. The price for each sweet should be different, and should be less than 25c. Write the suggested prices on the price tags. Underneath, draw a selection of coins that will allow you to buy two or three sweets, but not all of them. Work through the questions on photocopiable page 54 together for the particular combination of prices and coins you have chosen.
- Give out photocopiable page 54, and ask the learners to work individually or in pairs to complete it. They should make their own choices regarding the prices of the sweets, the coins they have, and what they buy. Ask the learners who finish to make up their own similar problems, and write them down.

- Write a money calculation on the board (for example 3 × 25c or 23c + 46c). Make up a number story to go with it (for example: *Lucy finds three quarters. How much money does she find?* or *Sulekha bought a raspberry chew for 23c and a packet of mints for 46c. How much did she spend altogether?*).
- Ask the learners to work in pairs, making up money calculations and number stories to go with them.

Plenary

- Ask the learners to share their number stories. Ask the other learners to say what calculation the story is based on, and then calculate the solution to the problem.

Success criteria

Ask the learners:

- Tell me a number story you've written.
- What is the calculation your number story is based on?
- What is the answer to that calculation?
- Explain how you worked it out.

Ideas for differentiation

Support: In the Starter, these learners could work in pairs with other learners. They may need some help choosing a selection of coins in the main activity.

Extension: Encourage these learners to extend their calculations to totals beyond $1 (for example to $5). You may need to revise dollar and cent notation with them (for example $2.47).

Name: _____

Money problems

These sweets are for sale in a sweet shop.
Write the price of each sweet on its price label.
The price of each sweet should be different, and should be less than 25c.

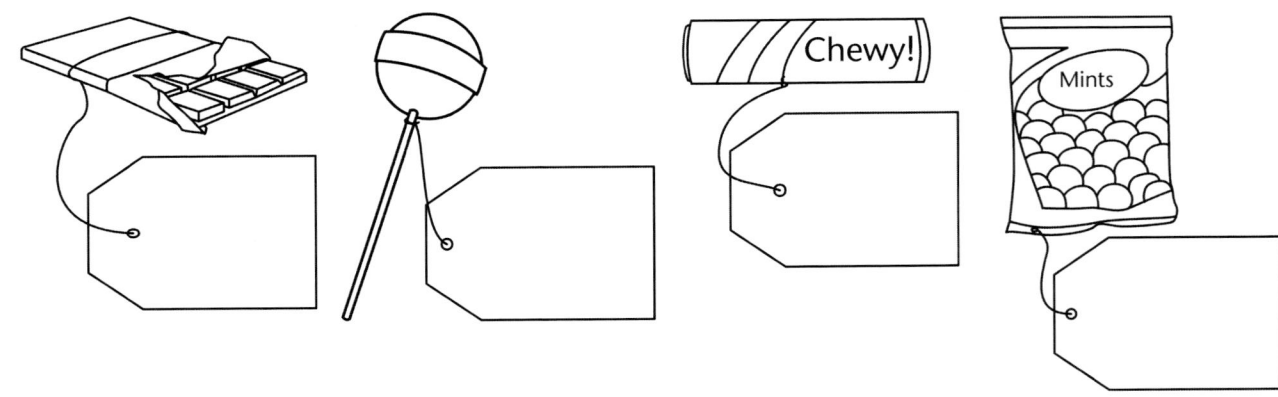

Draw the coins that you have:

1. What sweets can you buy with your coins? _____

_____

2. How much will the sweets cost? _____

_____

3. Which coins will you use to pay for the sweets? _____

_____

4. What change will you get? _____

_____

Cambridge Primary: Ready to Go Lessons for Maths Stage 3 © Hodder & Stoughton Ltd 2013

Length 1

- Choose and use appropriate units and equipment to estimate, measure and record measurements. (3Ml1)
- Use a ruler to draw and measure lines to the nearest centimetre. (3Ml4)
- Begin to understand everyday systems of measurement in length, weight, capacity and time and use these to make measurements as appropriate. (3Pt2)

Resources

Metre sticks; photocopiable page 56; rulers; tape measures; plain paper.

Starter

- Briefly revise the centimetre, metre and kilometre, and discuss the relationships between them. Show 1 cm and 1 m on a metre stick, and describe a distance of approximately 1 km (for example the distance between school and a particular landmark).
- Enlarge photocopiable page 56 onto A3 card and show the length cards one at a time. Ask: *Which unit would you use to measure this – kilometres, metres or centimetres?*
- Give the learners a unit of measurement (for example metres) and ask them to suggest things they would measure using that unit. Repeat for kilometres and centimetres.

Main activities

- Demonstrate how to use the ruler, the metre stick and the tape measure to measure the dimensions of objects to the nearest centimetre.
- Organise the learners into pairs and give each pair a ruler, a metre stick and a tape measure. Ask pairs to choose about eight objects to measure. For each measurement, the learners must make and record an estimate, then measure and record the actual length to the nearest centimetre, and finally calculate and record the difference between the two.

- On the board draw a rough sketch of a vegetable garden (or any plan based on a large rectangle with several smaller rectangular features inside it). Label each line on the plan with its 'real' dimensions in metres (leave out the 'm'; just write the number). Ask the learners to draw a scale drawing of the vegetable garden on plain paper, substituting centimetres for metres.

Plenary

- Ask the learners to look at the differences between their estimates and the actual measurements. Ask: *Did your estimates get better as you went along? Why do you think this is?*
- Ask the learners to check each other's scale drawings by measuring the lines.

Success criteria

Ask the learners:

- What unit would you use to measure the distance around your friend's head?
- What equipment would you use to measure the distance around your friend's head?
- Can you estimate the distance around your friend's head?
- What is the distance around your friend's head?

Ideas for differentiation

Support: In the main activity, pair these learners with a partner of average ability.

Extension: Challenge these learners to begin to estimate, measure and record lengths between 1 m and 2 m.

Which unit of length?

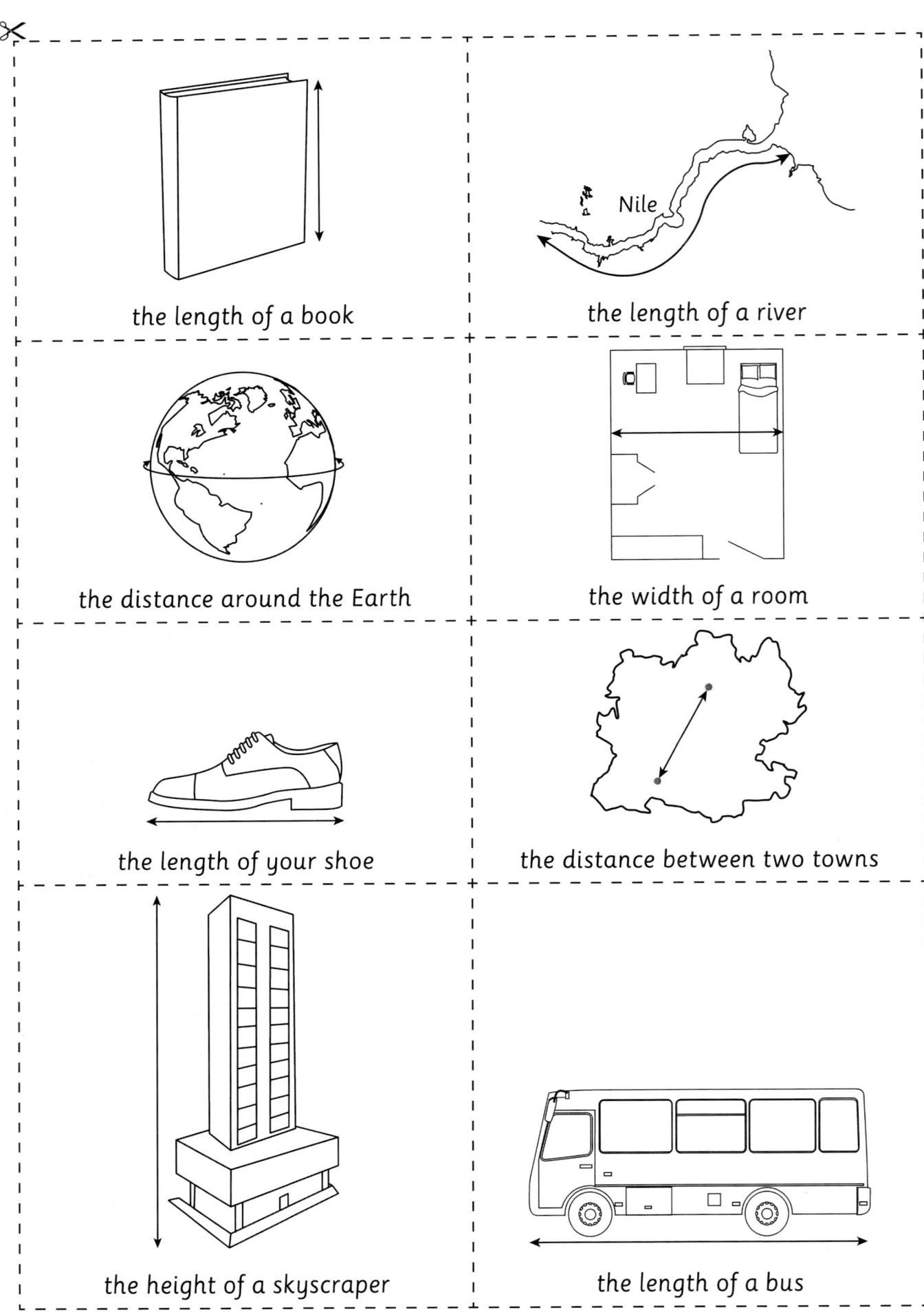

the length of a book

the length of a river

the distance around the Earth

the width of a room

the length of your shoe

the distance between two towns

the height of a skyscraper

the length of a bus

Mass 1

Learning objectives

- Know the relationship between kilometres and metres, metres and centimetres, kilograms and grams, litres and millilitres. (3Ml2)
- Read to the nearest division or half division, use scales that are numbered or partially numbered. (3Ml3)
- Solve word problems involving measures. (3Ml5)

Resources

Two paperclips; 1 kg bag of sugar or flour; pictures of about a dozen objects with a wide variety of masses (including some objects whose mass is typically measured in grams, such as a pencil, an apple, a mug or a sheet of paper, and some objects whose mass is typically measured in kilograms, such as a suitcase, a child, a car or an elephant); sticky tack; pan balances; spring scales; kitchen scales; bathroom scales; photocopiable page 58.

Starter

- Briefly revise the gram and the kilogram and the relationship between them. (We use two main units to measure mass: grams and kilograms. The gram is the smaller unit. One thousand grams equals 1 kilogram.) Explain that two paperclips have a mass of about 1 gram and a large bag of sugar or flour has a mass of about 1 kilogram. Ask a volunteer to compare these masses directly by holding one in each hand.
- Show the pictures one at a time. Ask: *Which unit would you use to measure the mass of this object?*
- Display the pictures on the board in a random order, and work together with the learners to order them from lightest to heaviest.

Main activities

- Look at the various instruments for measuring mass. For each instrument, tell the learners what the scale goes up to, and ask them to suggest a suitable object to find the mass of.

- Ask the learners to estimate the object's mass. Demonstrate how to use the instrument to find the mass of the object.
- Divide the class in half.
  - Ask one half of the class to work in pairs to choose objects, then estimate and measure their masses. Work with this group. Ask the learners to record their estimates and measurements in any way they choose. Observe the learners as they take measurements to ensure they are interpreting scales correctly.
  - Give the other half of the class copies of photocopiable page 58. Ask the learners to work in pairs to solve the word problems, recording their methods.
- Swap the groups over.

Plenary

- Read through one or two of the word problems. Ask the learners to give the answers, and explain how they worked them out.
- Ask: *Which of the measuring instruments was easiest / hardest to use? Why?*

Success criteria

Ask the learners:

- What unit would you use to measure the mass of your exercise book?
- What is the mass of your exercise book?
- Can you explain how you solved this word problem?

Ideas for differentiation

Support: The word problems on photocopiable page 58 are arranged from easiest to hardest; it may be appropriate for these learners to attempt only the first four questions.

Extension: These learners could work individually on the word problems and could devise their own problems to give to a friend.

Name: _____

Mass problems

1. Amina's bike has a mass of 15 kg.
   Leila's has a mass of 9 kg.
   What is the difference in the bikes' masses?

   _____

   _____

2. Sofia's mass is 33 kg and Mira's mass is 26 kg.
   What is their total mass?

   _____

   _____

3. A bag contains 500 g of sugar. Aisha uses 150 g of sugar.
   How many grams of sugar are left?

   _____

4. Stefan buys four bags of sand. Each bag has a mass of 5 kg.
   What is the total mass of sand?

   _____

5. Five packets of crisps have a total mass of 130 g.
   What is the mass of each packet?

   _____

6. In the cupboard there are three packets of cornflakes, each with
   a mass of 325 g.

   a)   What is the total mass of cornflakes? _____

   b)   How many more grams of cornflakes are needed to make 1 kg?

   _____

   _____

9. The Nasser family have 120 kg of luggage, equally divided between
   four suitcases. What is the mass of luggage in each suitcase?

   _____

Cambridge Primary: Ready to Go Lessons for Maths Stage 3 © Hodder & Stoughton Ltd 2013

Capacity 1

Starter

- Briefly revise the millilitre and the litre and the relationship between them. Hold up the teaspoon and the jug. Explain that the teaspoon has got 1 millilitre of water in it and that the jug has got 1 litre of water in it.
- Show the pictures one at a time. Ask: *Which unit would you use to measure the capacity of this container / the volume of this liquid?*
- Display the pictures on the board in a random order, and work together with the learners to order them from lowest to highest capacity / volume.

Main activities

- Display a copy of photocopiable page 60. Work through the first few questions together, discussing how to interpret each scale.

- Hold up two containers (for example a bowl and a cup). Ask: *How much do these containers hold together?* Model finding the answer by a) estimating the capacity of each container, b) measuring both capacities and referring back to the estimates to make sure the measurements are reasonable, c) doing a rough calculation using approximation, d) performing the actual calculation and e) checking the result against the approximation.
- On the board write calculations using pictures to represent containers, for example:

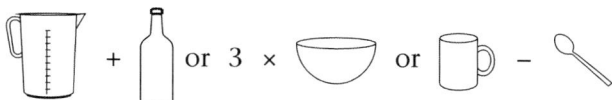

Ask half of the class to complete the calculations, using the method outlined above. Ask the other half of the class to complete photocopiable page 60. Swap over halfway through.

Plenary

- Work through the answers to photocopiable page 60.
- Regarding the practical measuring activity, ask: *Were there any containers whose capacities were difficult to measure?* (For example the dessert spoon.) *Why?* (Because its capacity is so small.) *How did you solve this problem?*

Name: _____

How many millilitres?

Read the scales. Write the number of millilitres of liquid in each container.

1.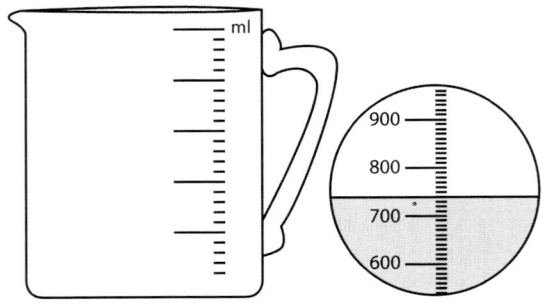

_____

2.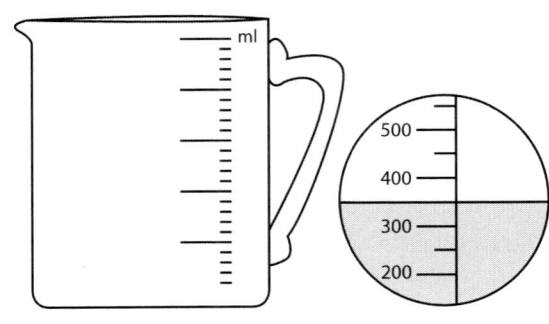

_____

3.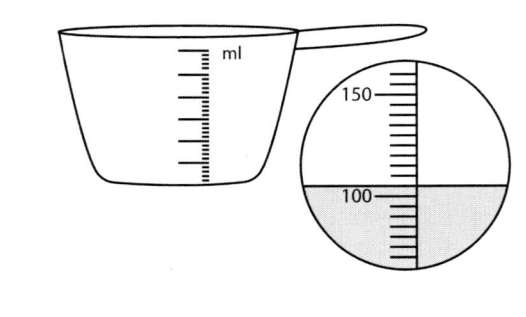

_____

4.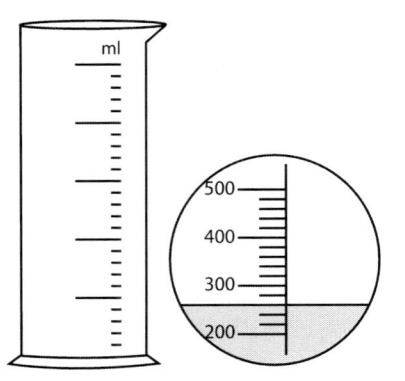

5.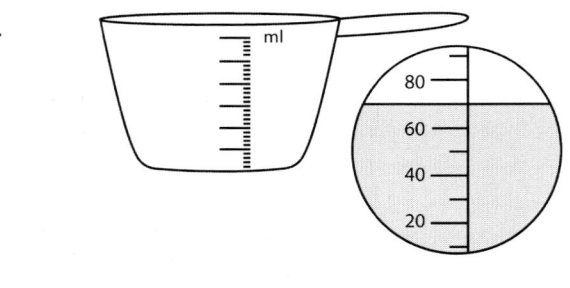

_____

6.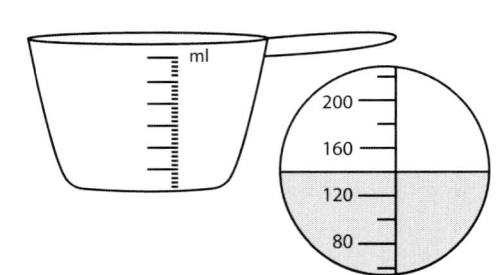

_____

7.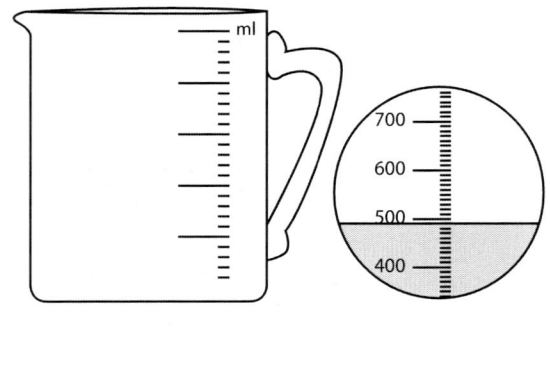

_____

8.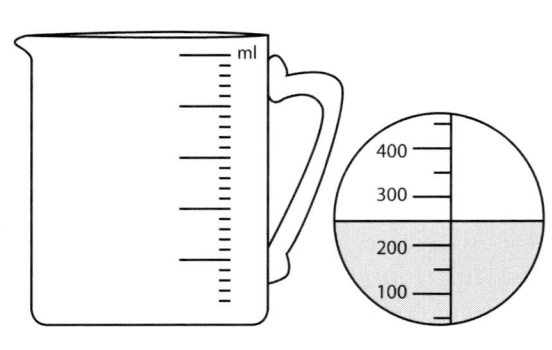

_____

Time 1

● Choose and use appropriate units and equipment to estimate, measure and record measurements. (3Ml1)

● Suggest and use suitable units to measure time and know the relationships between them. (3Mt1)

● Begin to understand everyday systems of measurement in length, weight, capacity and time and use these to make measurements as appropriate. (3Pt2)

Nine large cards each with a unit of time written on it (second, minute, hour, day, week, month, year, decade, century); electronic timers that record time to the nearest second; photocopiable page 62; electronic timers that record time to the nearest tenth or hundredth of a second; pencils; plain paper.

Starter

• Display the units of time cards in a random order, asking the learners to order them from shortest to longest.

• Ask the learners to define each unit of time (with the exception of the second) in relation to another unit (for example 1 minute = 60 seconds). Discuss the fact that 1 month = 28 to 31 days, and that 1 year = 365 or 366 days.

• Working in pairs, ask the learners to suggest an event whose duration it would be appropriate to measure using each unit.

Main activities

• Call for a volunteer who is wearing shoes with shoelaces. Ask: *How long do you think it will take Amir to untie his shoelaces and tie them back up again?* Record a range of estimates. Use and comment on all four standard ways of writing minutes and seconds: minutes / seconds; min / sec; m / s; ' / ".

• Demonstrate how to use the electronic timer, and then ask another volunteer to time the first volunteer doing the shoelace activity, and record this on the board.

• Compare the actual time the activity took to the various estimates. Ask: *Which estimate was the closest? How did you work it out?*

• Enlarge photocopiable page 62 onto A3 card and cut out to make 'Time me!' cards. Organise the learners into pairs, giving each pair a set of cards and a timer that records time to the nearest second. One learner should take a card and then perform the activity on it while their partner times them. Provide paper and pencils and ask the learners to draw a table (see below) to record estimates, measurements and differences:

Name	Activity	Estimate	Measure-ment	Differ-ence

Plenary

• Ask: *Was it easy to estimate the time the activities would take? Did your estimates get better as you got more practice?*

• Discuss how to find the difference between two times, one expressed in seconds and one in minutes and seconds.

Ask the learners:

● What unit or units of time would you use to measure how long a movie lasts?

● How many minutes are there in an hour?

● About how long will it take your friend to do the activity on this card?

● Time him / her! How long did it take? Record the time it took.

Support: Watch these learners using the timers, to ensure they are operating them correctly. Provide support with calculating differences between estimates and actual times.

Extension: Challenge these learners to begin using timers that record time to the nearest tenth or hundredth of a second.

'Time me!' cards

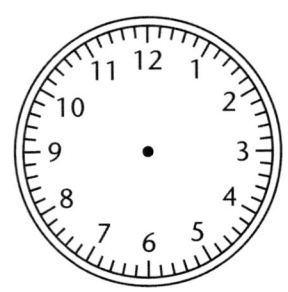

Count the pages in your exercise book.	Write your name 20 times.	Count aloud backwards from 50 to 0.
Recite the alphabet.	Choose a colour. Collect four different things in that colour.	Draw a house and a tree.
Recite the 10 times table.	Sing a verse of a song you both know.	Do 15 star jumps.
Make a paper aeroplane.	Write the 19th letter of the alphabet.	Hop 30 times.
Name 10 things beginning with t.	Write the name of the day that comes 10 days after Saturday.	Write the name of the month that comes 15 months after March.
Write your name in mirror writing.	Name five different rectangles in the room.	Choose a book. Write the 2nd word in the 3rd sentence in the 1st paragraph on the 5th page.

Time 2

Starter

- Make o'clock and half past times on the large standard clock face, asking the learners to say the time.

- Write a digital time to the hour or half hour on the board. Use digits like these:

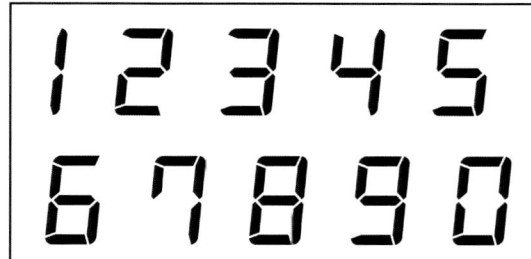

- Ask the learners to say the time, expressing half hours in two ways (for example 'seven thirty' or 'half past seven').

- In pairs, give each pair a small clock face, paper and pencils. Call out a time to the hour or half hour. One partner makes the time on the clock face and the other writes down the digital version.

Main activities

- Make a large teaching clock from photocopiable page 64, enlarging the page onto A3 card. Cut out the clock face and both clock hands. Colour the outer ring on the clock face and the long hand in one colour, and the inner circle on the clock face and the short hand in another colour. Attach the hands with a paper fastener.

- Use the clock to display times sequentially in five-minute intervals from five past the hour to half past the hour. (Do not go past half past the hour at this point.) Discuss the different ways of saying each time (for example 'ten minutes past five' or 'five ten') and demonstrate how to write it using digital notation (for example 5:10).

- Once the learners are confident with times up to half past, extend to times past the half hour to include 'to' times.

- Display random times on the large clock face, asking one learner to say the time, a second to say the same time in a different way, and a third to write the time as it would appear on a digital clock.

- Ask the learners to make their own clocks from A4 copies of photocopiable page 64 copied onto card, colouring the outer ring and long hand in one colour, and the inner circle and short hand in another colour. They should attach the hands to the clock face with a paper fastener.

Plenary

- Display the nine large digital time cards, arranged in a random order. Ask the learners to order the times shown on the cards from earliest to latest.

Teaching clock template

1. Cut out the pieces.

2. Colour the outer numbers on the clock face and the minute (long) hand in one colour. Colour the inner numbers on the clock face and the hour (short) hand in another colour.

3. Attach the hands to the clock face using a paper fastener.

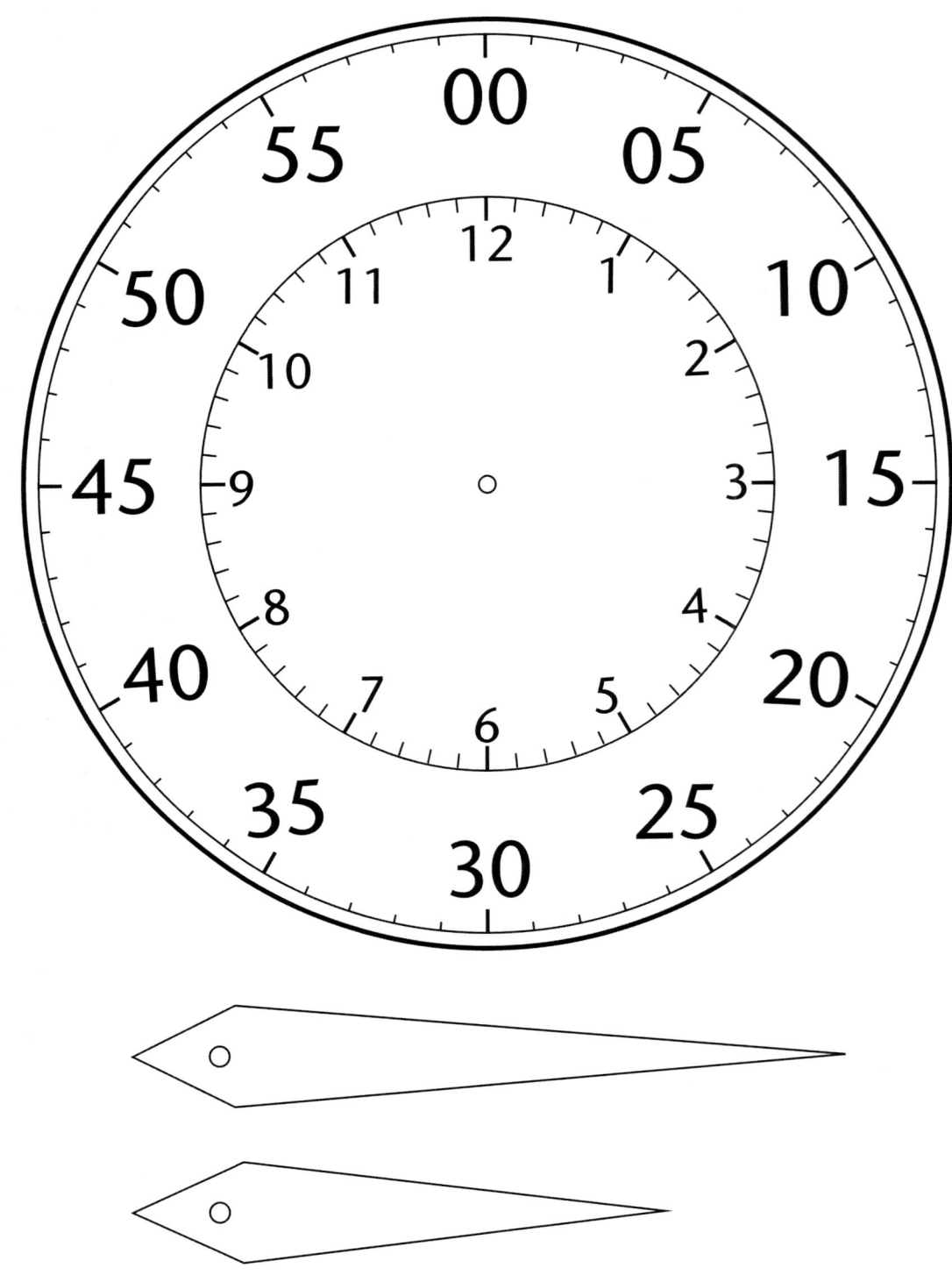

Cambridge Primary: Ready to Go Lessons for Maths Stage 3 © Hodder & Stoughton Ltd 2013

Time 3

Learning objectives

- Solve word problems involving measures. (3MI5)
- Explain a choice of calculation strategy and how the answer was worked out. (3Ps2)

Resources

One large clock made from photocopiable page 64; small versions of the same clock that the learners made in the previous lesson; photocopiable page 66.

Starter

- Revise telling the time in five-minute intervals on an analogue clock. Make a time on the large clock. Pick one learner to say the time, a second to say the time in a different way, and a third to write the time on the board as it would appear on a digital clock.
- Call out a time in five-minute intervals. Ask the learners to make the time on the clocks they made in the previous lesson.

Main activities

- Display a copy of photocopiable page 66. Read through the first problem.
- Ask: *What do you need to find out in order to solve this problem?*
- Ask the learners to perform the calculation (in pairs or individually) using whatever method they choose, and record their answers.
- Ask the learners to describe the methods they used to solve the problem (for example counting on in five-minute intervals on a timeline or doing the same around a clock face).
- Ask individuals or pairs to share their answers and explain how they worked them out.
- Repeat the process for a second problem.
- Organise the learners to work individually or in pairs, giving each individual or pair a copy of photocopiable page 66. For each problem, ask the learners to record an estimate, any written workings and the final answer.

Plenary

- Read through a few of the word problems that you have not already worked through together. Ask the learners to give the answers, and explain how they worked them out.
- Ask whether anyone else used a different method to solve the same problem.

Success criteria

Ask the learners:

- What did you need to do to solve this problem?
- What was your estimate? Explain how you worked it out.
- What answer did you get? Explain how you worked it out.
- Is your answer reasonable? How can you tell?

Ideas for differentiation

Support: Group these learners together and guide them through an extra problem. Ask them to complete only the first four problems, as these are the easiest.

Extension: Ask these learners to make up their own time problems and give them to a friend to solve.

Name: _____

Time problems 1

1. Ali puts biscuits in the oven at 3.30. They need to cook for 20 minutes.
    What time should Ali take the biscuits out of the oven?

    _____

2. Preeti watches a TV programme that is 35 minutes long.
    The programme starts at half past seven.
    What time does it end?

    _____

3. Johan and Diego are going to karate, which starts at 6.45.
    The journey takes 25 minutes. What is the latest they should leave home?

    _____

4. Malik starts playing a computer game at ten to eight.
    He stops at 8.45. How long did he play for?

    _____

5. Lunch is at 12.15. Lottie looks at her watch. It's 11.25.
    How long does she have to wait until lunch?

    _____

6. You arrived at a concert at 6.35. Your friend arrived at a quarter past seven.
    How long were you waiting for your friend?

    _____

7. You want to play the guitar for three quarters of an hour before you start
    watching TV at 5.35. What time must you start playing the guitar?

    _____

8. You get up at twenty past seven.
    It takes you 45 minutes to get ready to leave for school.
    At what time are you ready to leave for school?

    _____

Cambridge Primary: Ready to Go Lessons for Maths Stage 3 © Hodder & Stoughton Ltd 2013

Unit assessment

- Can you make up a number story to go with this calculation: $16 × 4? Solve the problem. How did you work out the answer?
- Can you estimate how long it will take you to write the alphabet forwards once and backwards once? Ask a friend to time you doing this. Compare your estimate to the measurement. How close was your estimate?

- What are the three main units of length? Explain the relationships between them.
- What units would you use to measure:
  a) the length of a swimming pool
  b) the distance between London and Dubai
  c) your hand span?

Summative assessment activities

Observe the learners while they are playing these games. You will quickly be able to identify those who appear to be confident and those who may need additional support.

Finding change from $1

This game assesses the learners' ability to find change from $1.

You will need:

A familiar board game (for example Ludo); ten-sided dice (0 to 9).

What to do

- Organise the learners into groups of four. Ask the groups to play the board game using the usual rules, but on each player's turn, they must answer a question before moving their playing piece.
- To generate the question, the player whose turn it is rolls two ten-sided dice marked 0 to 9, to make a two-digit number. This number represents an amount of money in cents. The player must say how much change they would get from $1 if they were to buy an item of that price. The rest of the players also do the calculation, so that they can check that the player whose turn it is has got the right answer.

Mass estimating and measuring game

This game assesses the learners' ability to estimate, measure and record mass.

You will need:

Instruments to measure mass.

What to do

- Organise the learners into groups of six. Give each group a range of mass-measuring instruments.
- Ask each player to choose one object from the classroom. Each player writes the names of the six chosen objects from lightest to heaviest, then estimate the mass of each object and record their estimates on their ordered list.
- Each player must find and record the mass of the object they chose. They should then compare each player's estimate with this actual mass, giving one point to whichever player has the closest estimate. The player with the most points wins.

Distribute photocopiable page 68. Ask the learners to read the questions and write the answers. They should work independently.

Name: _____

Looking at measures

1. Use a ruler to draw a line that is 7 cm long.

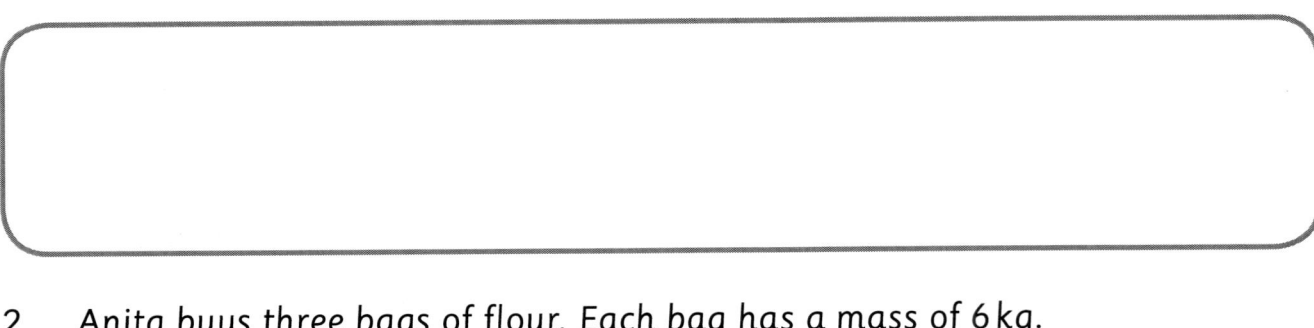

2. Anita buys three bags of flour. Each bag has a mass of 6 kg.
   What is the total mass of flour?

   _____

3. How much water is in the jug?

   _____

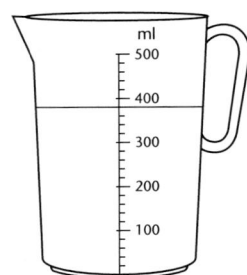

4. What is the time? Write the time in two ways.

   _____

   _____

5. Home time is at 3.10. Sara looks at her watch. It's 2.45.
   How long does she have to wait until home time?

   _____

6. Which instrument would you use to measure around your waist?

   a) ruler   b) metre stick   c) tape measure

   _____

7. How much money is this?

   _____

 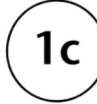    

Cambridge Primary: Ready to Go Lessons for Maths Stage 3 © Hodder & Stoughton Ltd 2013

Comparing and ordering numbers 1

Learning objectives

- Place a three-digit number on a number line marked off in multiples of 100. (3Nn9)
- Compare three-digit numbers, use < and > signs, and find a number in between. (3Nn11)
- Order two- and three-digit numbers. (3Nn12)

Resources

Two large sets and plenty of small sets of 0 to 9 digit cards; large 0 to 1000 number line marked off in multiples of 10 with the multiples of 100 labelled (e.g. a metre stick on which you have labelled every 10 cm as a multiple of 100); pencils; plain paper; photocopiable page 70.

Starter

- Display the 0 to 1000 number line. Shuffle together two sets of large 0 to 9 number cards. Call up two volunteers and ask each volunteer to draw three cards and use them to make the largest possible three-digit number. Each player should place their number on the number line then compare the numbers. The player who makes the larger number scores a point.
- Ask the learners to play the game in pairs. Give each pair two sets of small cards. The winner is the first player to reach five points. Finishers could replay the game, this time competing to make the smaller number.

Main activities

- Introduce and explain the symbols for 'is less than' and 'is greater than'. Compare the symbols to the jaws of a crocodile, opening up to eat the larger number. Write the following on the board. Ensure each empty box is big enough to hold a large number card:

☐ < ☐

☐ > ☐

A number between ☐ and ☐ is ☐.

- Make two three-digit numbers from the large 0 to 9 digit cards. Use these numbers to complete the number sentences on the board.
- Ask the learners to do the activity in pairs, using paper and pencils and the small number cards.
- Write half a dozen two- and three-digit numbers on the board, and ask the learners to work in pairs to order them from smallest to largest. Ask the learners to write their own list of unordered numbers, and give it to their partner to order.

Plenary

- Give groups of learners a set of cards made from photocopiable page 70. Challenge the groups to race to order the numbers.
- Ask: *Was the smallest number quick to find? Why? / Why not? What about the biggest number? Explain how you worked out the order of the numbers in between.*

Success criteria

Ask the learners:

- Is this number sentence correct? 586 < 568 Why? / Why not?
- Write a number sentence using the symbol >.
- Write a number between 393 and 397. Mark your number on a number line.
- Write these numbers in order from smallest to largest: 89, 54, 108, 180, 45, 316

Ideas for differentiation

Support: In the paired activity at the end of the starter, these learners could make two-digit instead of three-digit numbers.

Extension: In the final main activity, challenge these learners to write long lists of numbers for their partners to order, and to include some four-digit numbers.

Numbers to order

81	46	38
39	36	44
18	976	470
848	374	476
409	610	666

Place value 3

- Count on and back in ones, tens and hundreds from two- and three-digit numbers. (3Nn3)
- Understand what each digit represents in three-digit numbers and partition into hundreds, tens and units. (3Nn5)
- Find 1, 10, 100 more / less than two- and three-digit numbers. (3Nn6)
- Find 20, 30, ... 90, 100, 200, 300 more / less than three-digit numbers. (3Nc18)

Counting stick; pencils; plain paper; large display, and tabletop, place value cards (hundreds, tens and units); ten-sided dice made from photocopiable page 72.

Starter

- Use the counting stick to count on and back from two- and three-digit numbers in ones, tens and hundreds. Emphasise the changing digit in each number (for example 3**3**, 3**4**, 3**5** / 2**4**1, 2**5**1, 2**6**1 / 6**1**8, **7**18, **8**18).
- Ask the learners to find two-digit multiples of 10 more or less than two- or three-digit numbers, by counting on or back in tens, for example: *Find 40 more than 125 by counting on by 10 four times.* (135, 145, 155, 165.) Ask the learners to find three-digit multiples of 100 more or less than a two- or three-digit number by counting on or back in hundreds, for example: *Find 300 less than 832 by counting back by 100 three times.* (732, 632, 532.)

Main activities

- Organise the learners into pairs. Give each pair paper and pencils. Draw this place value chart on the board:

Hundreds	Tens	Units

- Write a two- or three-digit number in the chart. Ask the learners to write the number that is 1, 10 or 100 more or less. Ask: *Which digit did you need to change?* Include some examples that involve crossing the tens boundary (for example 1 more than 239) or crossing the hundreds boundary (for example 10 less than 704).
- Display a three-digit number made from place value cards (for example 673). Ask: *Which cards did I use to make this number?* Pull the cards apart and write 673 = 600 + 70 + 3. Repeat for other three-digit numbers, including some that use just two cards (for example 305 or 630).
- Give each pair of learners a ten-sided dice made from photocopiable page 72, and a set of small place value cards. Ask the learners to generate a three-digit number by throwing the dice three times, make the number using place value cards, and record the partitioning.

Plenary

- Write a partitioning on the board (for example 800 + 30), asking the learners to write the number (for example 830). Challenge the learners to work out the answer without using the place value cards. Repeat for other three-digit numbers.

Ask the learners:

- Count on in tens from 245.
- Write the number that is 100 less than 612. Explain how you worked out the answer.
- Partition 427.
- Write this number: 800 + 50.

Support: These learners could make two-digit numbers instead of three-digit numbers.

Extension: In the main activity, challenge these learners to partition numbers without first making them from place value cards. Extend to include four-digit numbers.

Templates for ten-sided dice

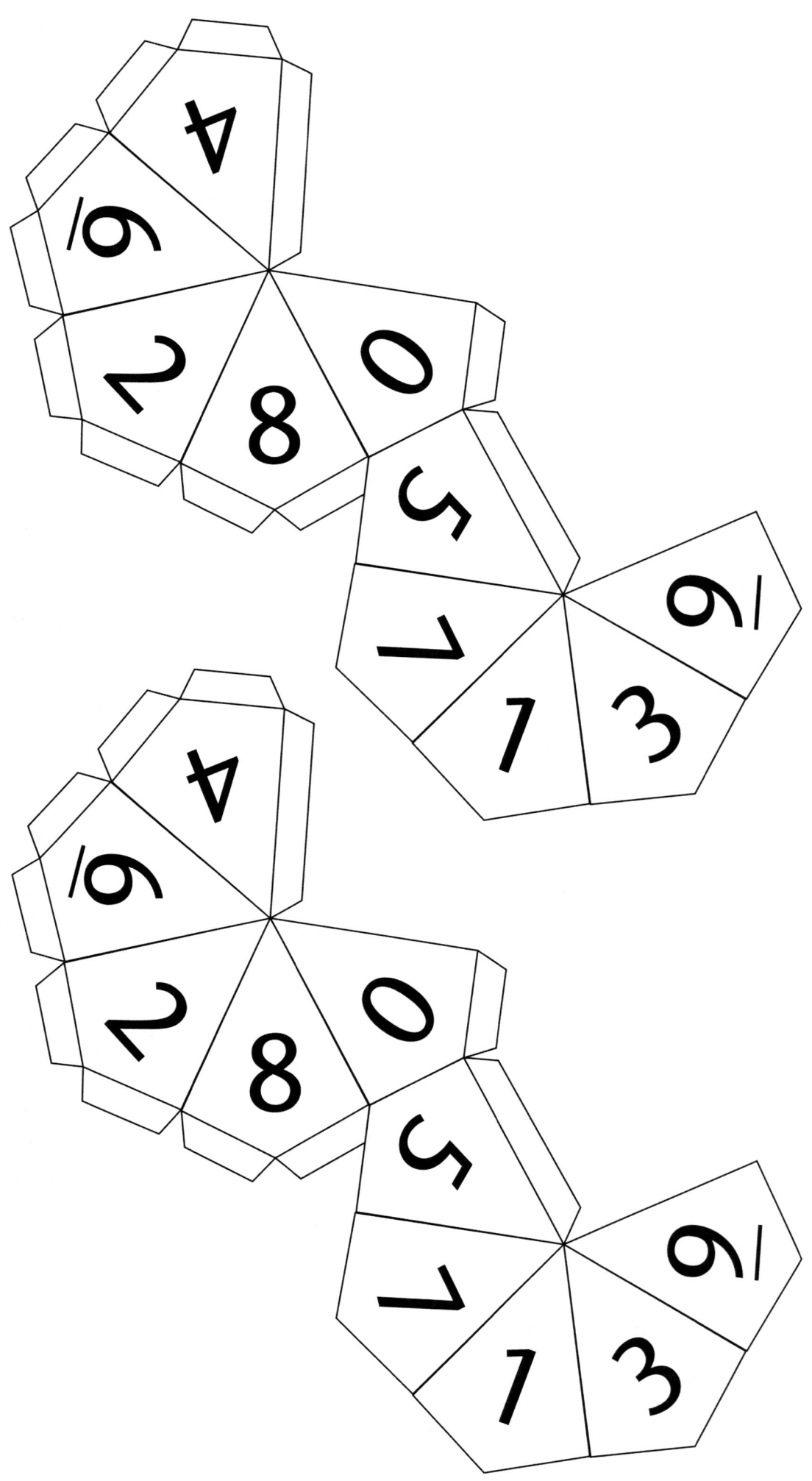

 Cambridge Primary: Ready to Go Lessons for Maths Stage 3 © Hodder & Stoughton Ltd 2013

Place value 4

Learning objectives

- Multiply two-digit numbers by 10 and understand the effect. (3Nn7)
- Investigate a simple general statement by finding examples which do or do not satisfy it. (3Ps8)

Resources

Two stacks of sticky notes in two different colours; laminated place value charts made from photocopiable page 74; dry wipe pens; cloths; pencils; erasers; calculators.

Starter

- Divide the class into two teams and sit each team in a line, one learner behind the other. At the front of each line place a pencil and a stack of sticky notes in a particular colour. Call out a number between 0 and 10. The learner at the front of the line must multiply the number by 10, write the product on a sticky note and race to give it to you.
- Keep the first correct answer you receive. Discard the other sticky note.
- When every learner has had a go, count the number of sticky notes you have in each colour. The team with the most wins.

Main activities

- Organise the learners into pairs and give each pair a laminated copy of a place value chart made from photocopiable page 74, a dry wipe pen and a cloth. Alternatively they can use a paper copy of photocopiable page 74, a pencil and an eraser.
- On the board draw a place value chart like the one on photocopiable page 74. Write a two-digit number in the chart, asking: *How can we multiply this number by 10?* Demonstrate moving each digit one place to the left and writing a 0 to fill the empty units column.

- Ask the learners to multiply various two-digit numbers by 10 using a place value chart.
- Ask the learners to use calculators to investigate the statement 'To multiply any number by 10, write a zero on the end'. After a few minutes, write a decimal number on the board, for example 2.5. Ask the learners to multiply this number by 10 on their calculators. Discuss the result. Challenge the learners to find other numbers that do not 'follow the rule'.

Plenary

- Discuss with the learners what they found out when investigating the statement about multiplying by 10. Revise the statement to 'To multiply any **whole** number by 10 put a zero on the end' and / or 'To multiply any number by 10 **move its digits one place to the left**'.

Success criteria

Ask the learners:

- Can you multiply 58 by 10?
- What happens to a number when you multiply it by 10?
- Can you give me an example of a number that you cannot multiply by 10 simply by putting a zero on the end?

Ideas for differentiation

Support: Give these learners instruction in using the calculator to multiply numbers by 10.

Extension: In the first main activity, ask these learners to extend the investigation by writing their own statements about multiplying by 100 and / or 1000.

Place value charts

Thousands	Hundreds	Tens	Units

Thousands	Hundreds	Tens	Units

Thousands	Hundreds	Tens	Units

Thousands	Hundreds	Tens	Units

Estimating and rounding 1

Learning objectives

- Round two-digit numbers to the nearest 10 and round three-digit numbers to the nearest 100. (3Nn8)
- Give a sensible estimate of a number as a range (e.g. 30 to 50) by grouping in tens. (3Nn13)

Resources

Pencils; plain paper; estimation cards made from photocopiable page 76; metre stick; jam jar full of jelly beans or other small sweets of uniform size (make a note of how many beans are in the jar); laminated number lines (with 10 unlabelled divisions).

Starter

- Ask: *What is an estimate? Give me an example of a situation in which you might want to make an estimate. Why would making an estimate be useful in this situation?*
- Give each learner several sheets of paper and a pencil. Make estimation cards using photocopiable page 76. Write the answer on the back of each card (see below) and sort them into set A and set B.
- Hold up the cards from set A one at a time, showing each card for about five seconds. Ask the learners to make and record an estimate for each card.
- Discuss their answers and methods of estimation. Repeat for set B.

    Answers: **A1** 23; **A2** 32; **A3** 21; **A4** 35; **A5** 28; **A6** 19; **B1** 13; **B2** 64; **B3** 320.

Main activities

- Use a metre stick to draw a number line on the board that is 1 m long and has divisions every 10 cm.

- Label the left end of the number line 20 and the right end 30. Ask a learner to mark 26 on the number line. Ask: *Which multiple of 10 is 26 nearer to: 20 or 30?* Explain that this is called 'rounding to the nearest 10'. Repeat for other two-digit numbers, relabelling the ends of the number line as appropriate. Teach the rounding up rule for numbers that come halfway.
- Write a two-digit number on the board. Ask the learners to sketch a number line showing the number and the two multiples of 10 it lies between. They circle the multiple of 10 that the number rounds to.
- Repeat the process in the previous two bullet points for rounding three-digit numbers to the nearest 100.

Plenary

- Display the jam jar full of jelly beans. Ask the learners to work in pairs to estimate the number of jelly beans in the jar. Collect estimates and discuss methods used.

Success criteria

Ask the learners:

- Estimate how many cars there are outside on the street / pencils there are in this pot. How did you reach your estimate?
- Give me a number that becomes 60 when you round it to the nearest 10.
- What is 443 rounded to the nearest 100?

Ideas for differentiation

Support: Provide these learners with laminated number lines for the third main activity.

Extension: Give these learners homework of devising their own estimation challenges to present to the class.

Estimation cards

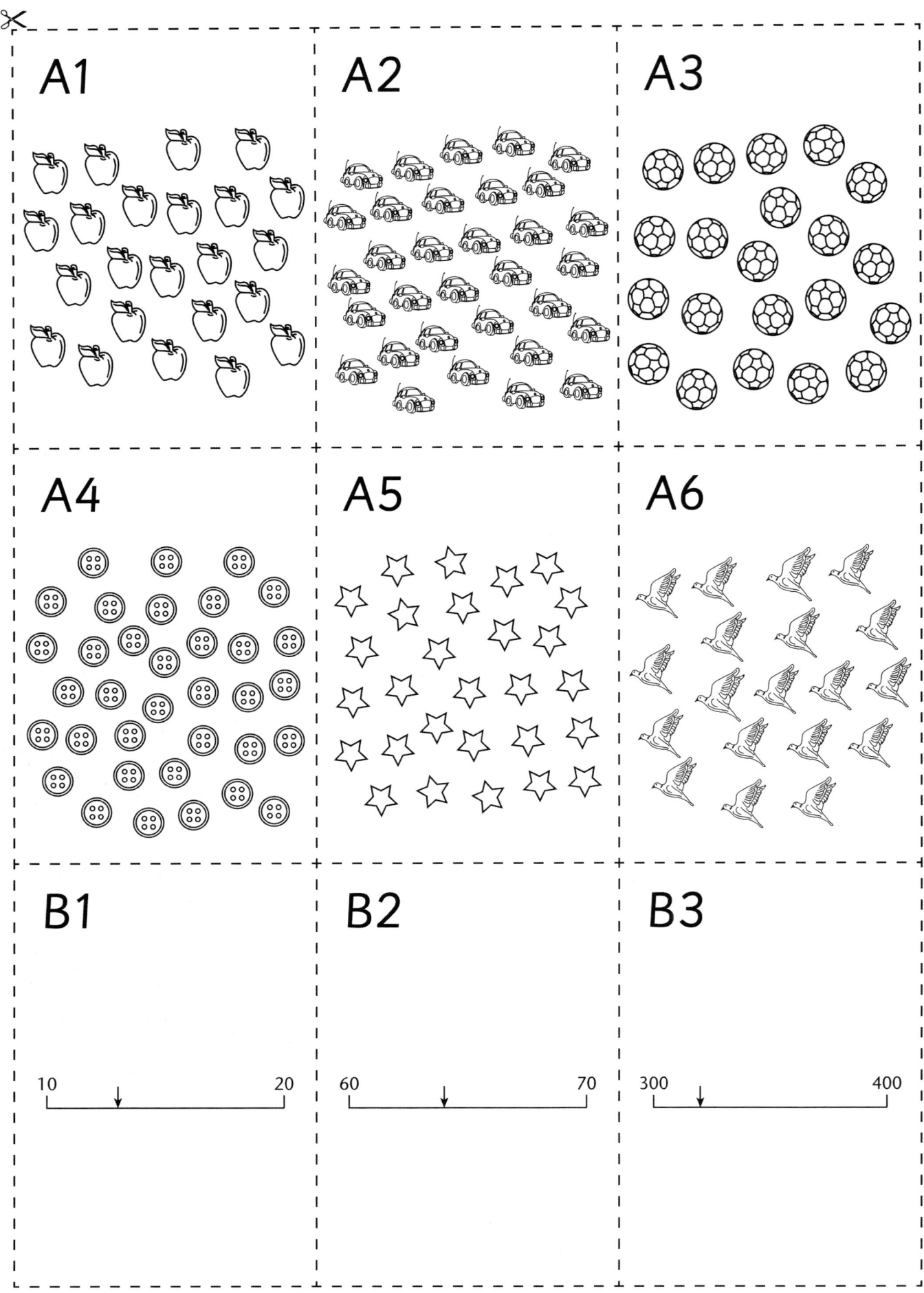

A1

A2

A3

A4

A5

A6

B1

B2

B3

10   ↓   20

60   ↓   70

300   ↓   400

Estimating and rounding 2

- Make sense of and solve word problems and begin to represent them. (3Pt3)
- Estimate and approximate when calculating, and check working. (3Pt10)
- Make a sensible estimate for the answer to a calculation, e.g. using rounding. (3Pt11)
- Explain a choice of calculation strategy and show how the answer was worked out. (3Ps2)
- Explain methods and reasoning orally, including initial thoughts about possible answers to a problem. (3Ps9)

Resources

Photocopiable page 78.

Starter

- Draw a 10-division number line on the board and revise using it to round a two-digit number to the nearest 10. Rub out the number line and give the learners another number to round. Ask them how they worked out the answer (look at the units digit, round down for less than 5, round up for equal to or greater than 5).
- Practise rounding measurements to the nearest 10 units. Write a measurement on the board (for example 43 kg). Ask the learners to round the measurement to the nearest 10 units (for example the nearest 10 kilograms) without using a number line. Include durations of time, amounts of money, distances, volumes and masses.

Main activities

- Display a copy of photocopiable page 78. Model solving one of the problems using these steps:
  1. Write down the calculation you need to do.
  2. Estimate the answer to the calculation by rounding the numbers (where helpful), and performing a quick mental calculation with the rounded numbers.
  3. Record your estimate.
  4. Perform the calculation.
  5. Check the answer by comparing it with your estimate.

- Choose another problem, asking the learners to perform one step at a time, discussing methods and results between each step. After step 4 ask the learners to describe the calculation strategy they are using and explain why they chose it. After step 5 ask: *What should you do if your answer is very different from your estimate?*
- Give out copies of photocopiable page 78. Ask the learners to work through the rest of the problems independently or in pairs.

Plenary

- Read through a couple of the problems that you didn't work through together during the main activity. Ask the learners to share their estimates and their final answers, and explain their methods.

Success criteria

Ask the learners:

- What calculation did you need to do in order to solve this problem?
- What was your estimate? How did you work it out?
- What is the answer to the problem? How did you work it out?
- Is your answer reasonable? How do you know?

Ideas for differentiation

Support: The problems on photocopiable page 78 are arranged from easiest to hardest; it may be appropriate for these learners to attempt only the first five questions.

Extension: These learners should work individually rather than with a partner. Early finishers could devise similar problems to give to a friend.

Name: _____

Airport problems

1. To get to the airport, the Chen family travels for 28 minutes
   on a train and 23 minutes in a taxi.
   How long does the journey take?

   _____

2. Lin and Huan have $100 to spend on holiday.
   They spend $39 in the shops at the airport.
   How much money do they have left?

   _____

3. The Chens' suitcases have masses of 31 kg, 27 kg, 18 kg and 23 kg.
   What is their total mass?

   _____

4. At the airport Mr Chen buys four umbrellas for $19 each.
   How much does he pay?

   _____

5. At the shop Mrs Chen buys four 48 ml bottles of perfume.
   How many millilitres of perfume does she buy?

   _____

6. Mrs Chen's perfume bottles each have a mass of 152 g.
   What is their total mass?

   _____

7. The Chens take three flights: one of 103 km, one of 197 km and one of 308 km.
   How far do they fly altogether?

   _____

8. During the flights, Lin and Huan watch two films. 'Hamster of Doom' is
   1 hour 27 minutes long, and 'Funny Friday' is 1 hour 24 minutes long.
   How long do Lin and Huan spend watching films?

   _____

    Cambridge Primary: Ready to Go Lessons for Maths Stage 3 © Hodder & Stoughton Ltd 2013

Addition and subtraction facts 3

- Count on and back in steps of 2, 3, 4 and 5 to at least 50. (3Nn4)
- Know the following addition and subtraction facts: multiples of 100 with a total of 1000; multiples of 5 with a total of 100. (3Nc2)
- Explore and solve number problems and puzzles. (3Ps3)

1 to 100 square; counting stick; cards made from photocopiable page 80.

Starter

- Display a 1 to 100 square. Choose a starting number that is an odd number between 1 and 29. Lead the class in counting on from your starting number in twos to at least 51, and back to the starting number. Point to each number on the hundred square as you say it. Next, count on and back from the same starting number using steps of the same size, but using a counting stick instead of the hundred square.
- Repeat the activity for counting on and back in steps of 3, 4 and 5. Choose a starting number between 1 and 30 that is not a multiple of the number you will be counting in.

Main activities

- On the board draw a number line from 0 to 100 labelled with multiples of 5. Circle one of the multiples of 5. Ask: *How many more to make one hundred?*
- Repeat the previous activity, drawing a number line from 0 to 1000 labelled with multiples of 100. Ask: *How many more to make one thousand?*
- Enlarge photocopiable page 80 onto A3 card to make a large set of number cards. Shuffle the cards and display them face down in a grid of six rows of five cards. Play a memory game, with players taking it in turns to turn over two cards. Players who turn over a pair of numbers totalling 100 or 1000 keep the cards. The winner is the player with the most cards when there are no cards left in the grid.

- Draw a three by three square grid on the board. Underneath the grid write the following numbers: 10, 20, 30, 40, 50, 60, 70, 80, 90. Ask the learners: *Can you arrange these numbers in the grid, so that each row, column and diagonal adds up to 150?*

Plenary

- Discuss puzzle solutions, and ask the learners to explain their methods. There are various ways of solving the puzzle, but each correct solution has 50 in the centre square. Here is one correct solution:

20	70	60
90	50	10
40	30	80

Ask the learners:

- Write two multiples of 5 that total 100.
- Write two multiples of 100 that total 1000.
- Which number comes next? 30, 27, 24, 21, ...
- Write the first six multiples of 4.

Support: In the memory game, give these learners just the multiples of 5 cards (remove the multiples of 100 cards).

Extension: After they have played the memory game once, ask these learners to devise their own game using the cards.

Memory game

5	10	15	20
25	30	35	40
45	50	50	55
60	65	70	75
80	85	90	95
100	200	300	400
500	500	600	700
800	900		

 Cambridge Primary: Ready to Go Lessons for Maths Stage 3 © Hodder & Stoughton Ltd 2013

Addition and subtraction strategies 1

- Add and subtract pairs of two-digit numbers. (3Nc14)
- Choose appropriate mental strategies to carry out calculations. (3Pt1)
- Check the results of adding two numbers using subtraction, and several numbers by adding in a different order. (3Pt4)
- Check subtraction by adding the answer to the smaller number in the original calculation. (3Pt5)
- Consider whether an answer is reasonable. (3Pt12)

One large and plenty of small 0 to 9 number fans (see photocopiable page 24 for template); 0 to 9 ten-sided dice (see photocopiable page 72 for template); board games; spinners made from photocopiable page 82; calculators; 0 to 5 spinners.

Starter

- Give each learner a small 0 to 9 number fan. Show two single-digit numbers on the large 0 to 9 number fan. Ask: *What is the total of these two numbers?* Ask the learners to show the total on their number fan. (Do not show pairs of numbers that total 11, as this number cannot be made on the number fans.) Keep the pace brisk.
- Repeat the activity, asking the learners to find the difference between the two numbers.

Main activities

- Ask a volunteer to roll a 0 to 9 ten-sided dice four times to generate two two-digit numbers, and then write the numbers on the board.
- Ask the learners to find the total of the two numbers. Discuss the range of strategies used (for example partitioning and recombining, counting up on a number line, and so on). Briefly outline any common mental-with-jottings strategies that the learners fail to mention. Ask: *How do we know whether the answer is reasonable? How could we check the answer?*

- Using the same numbers, ask the learners to find the difference between them. Once again, discuss calculation strategies and checking methods.
- Organise the learners into groups of four. Give each group a ten-sided dice, a + / – spinner made from photocopiable page 82 and a calculator. Each player should take it in turn to roll the dice four times to generate two two-digit numbers, then spin the spinner to determine whether to find the total or the difference. The player who rolled the dice does the calculation using whatever method they like, and then gives the answer. Another player checks their answer using the calculator. A player gets a point for every correct answer. The winner is the player with the most points when every player has had a pre-decided number of turns.

Plenary

- Write these numbers on the board: 8, 6, 8, 5, 12, 4.
- Ask the learners to find the total and check their answer by adding in a different order.
- Discuss strategies (for example finding pairs that make 10 or 20, looking for doubles and near doubles, starting with the largest number).

Ask the learners:

- Explain how you would find the total of 29 and 45.
- Find the difference between 82 and 56. How did you work it out?
- Find the total of 7, 6, 15, 3, 6 and 5. Do you think your answer is reasonable? How could you check your answer?

Support: In the game, give these learners a 0 to 5 spinner instead of a ten-sided dice. This will keep totals within 100.

Extension: In the game, whenever these learners spin an addition, they should roll the dice twice more to generate an extra two-digit number to add on.

Spinner template

Copy onto card and cut out. Place the point of a pencil in the middle of the spinner with a paper clip underneath it. Spin the paper clip and see where it lands.

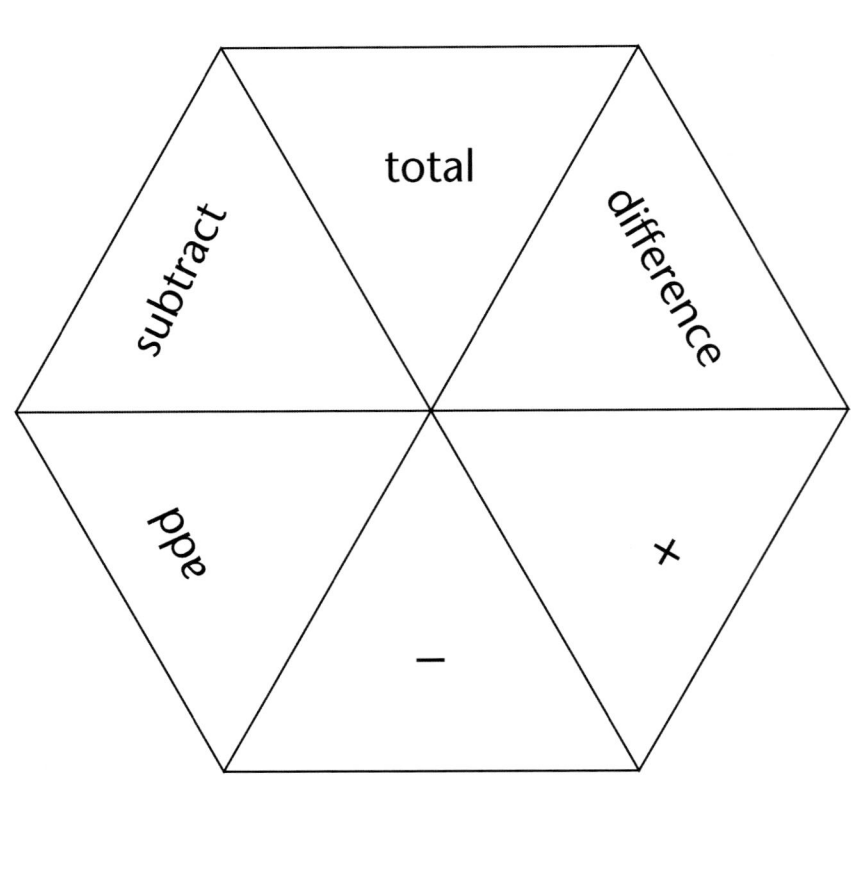

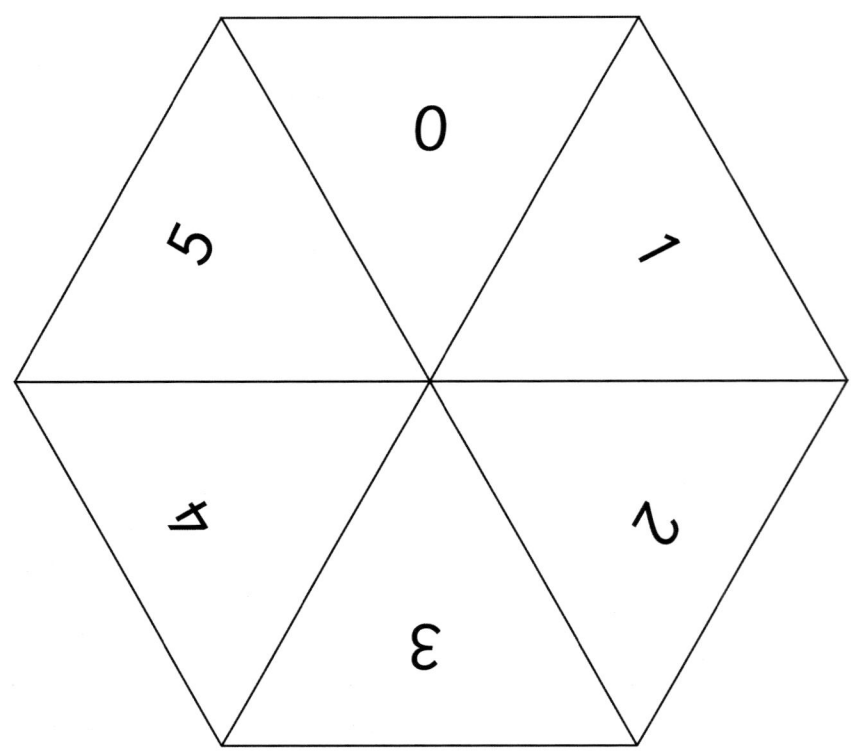

Addition and subtraction strategies 2

Learning objectives

● Add three-digit and two-digit numbers using notes to support. (3Nc15)

● Add / subtract single-digit numbers to / from three-digit numbers. (3Nc17)

● Make up a number story to go with a calculation. (3Ps1)

Resources

Cards made from photocopiable page 84; timer; number lines or place value blocks.

Starter

- Revise addition and subtraction facts up to 20 by playing a follow-me card game:
  - Enlarge photocopiable page 84 onto A3 card and cut out the cards. Find and keep the START card. Give out the rest of the cards. (Some learners may need to share one card between two.) Read the START card aloud. The learner who has the answer to that calculation written on their card should read their card aloud, and so on, until the END card is reached.
  - Play a few more rounds, redistributing the cards between rounds. Time the activity, and challenge the class to beat their previous best time.

Main activities

- On the board write an addition involving a three-digit number and a single-digit number (for example 476 + 7).

- Demonstrate how to perform the calculation using an informal (sketched) number line. Mark the three-digit number on the number line and then count up aloud in ones (for example saying: *477, 478, 479, 480, 481, 482, 483*), and draw a curve for each jump, marked '+1', until you have made seven jumps. Record the answer in a number sentence (for example 476 + 7 = 483).

- Give the learners a couple of similar calculations to perform themselves, using the same method (for example 524 + 8 or 308 + 6), and then challenge them to perform a subtraction using

the same method (for example 463 – 5 or 121 – 9). For each calculation, ask the learners to make up a number story to go with it.

- On the board write an addition involving a three-digit number and a two-digit number (for example 209 + 55). Demonstrate how to perform the calculation using an informal number line, using annotated curved jumps as before; first counting on in tens (for example to 219, 229, 239, 249 and 259), and then counting on in ones (for example 260, 261, 262, 263, 264).

- Give the learners a couple of similar calculations to perform themselves, using the same method (for example 156 + 28 or 765 + 34), and then challenge them to perform a subtraction using the same method (for example 243 – 35 or 262 – 48). For each calculation, ask the learners to make up a number story to go with it.

- Ask the learners who finish the calculations you have given them to make up similar calculations of their own and write number stories to go with them.

Plenary

- Ask selected learners to share the number stories they have written, asking the rest of the class to perform the calculations.

Success criteria

Ask the learners:

● Find the total of 236 and 58. Explain how you worked it out.

● Calculate 601 – 7. Explain how you worked it out.

● Can you make up a number story to go with this calculation?

Ideas for differentiation

Support: Provide these learners with equipment to support their calculations (for example number lines or place value blocks).

Extension: In the starter activity, if there are any cards left over once everyone has a card, give these learners an extra card.

Follow-me cards

START	10	2
4 + 6 =	13 − 11 =	9 − 3 =
6	16	11
8 + 8 =	3 + 8 =	4 + 9 =
13	1	18
11 − 10 =	9 + 9 =	14 − 5 =
9	19	7
9 + 10 =	15 − 8 =	6 + 11 =
17	3	14
12 − 9 =	8 + 6 =	13 − 9 =
4	15	8
7 + 8 =	13 − 5 =	5 + 7 =
12	5	0
18 − 13 =	19 − 19 =	END

Multiplication and division facts 1

Learning objectives

- Know multiplication / division facts for 2×, 3×, 5× and 10× tables. (3Nc3)
- Understand the relationship between multiplication and division and write connected facts. (3Nc26)
- Check multiplication by reversing the order, e.g. checking that 6 × 4 = 24 by doing 4 × 6. (3Pt6)
- Describe and continue patterns which count on or back in steps of 2, 3, 4, 5, 10 or 100. (3Ps5)

Resources

Pencils; plain paper; cards made from photocopiable page 86; times table charts for the 2, 3, 5 and 10 times tables.

Starter

- Give each learner several sheets of paper and a pencil. On the board write a number pattern that counts on or back in constant steps of 2, 3, 4, 5 or 10. When counting in 2s and 10s include patterns containing numbers that are not multiples of the step size, for example 7, 9, 11 ... (not multiples of 2) or 74, 64, 54, ... (not multiples of 10).
- Ask the learners to write the next three numbers in the pattern, and then describe the pattern (for example 'adding four' / 'counting back in 5s' / '3 less each time').

Main activities

- Enlarge photocopiable page 86 onto A3 card and cut out the array cards. Display one of the cards and explain that this type of rectangular grid is called an array. Ask: *How many objects does the array contain? How did you work it out?* (Hopefully, some will have used multiplication.)

- Work together to derive the multiplication facts shown in the array (for example 3 × 5 = 15 and 5 × 3 = 15) and the corresponding division facts (for example 15 ÷ 5 = 3 and 15 ÷ 3 = 5). Show more cards, asking the learners to derive the families of facts and record them.
- Display charts for the 2, 3, 5 and 10 times tables. Divide the learners into four groups. Ask each group to draw arrays to model facts from a different times table, and write the corresponding family of facts underneath each array.

Plenary

- Ask volunteers to share the arrays they have drawn and the corresponding families of facts they have written.
- Display the times table charts for the 2, 3, 5 and 10 times tables and practise chanting them.

Success criteria

Ask the learners:

- Write the next three numbers in this pattern: 12, 15, 18, ...
- Describe this number pattern: 78, 68, 58, 48
- Draw an array to show 21 ÷ 3. What's the answer?
- (Showing an array card:) Write the family of facts for this array.

Ideas for differentiation

Support: In the second main activity, give these learners more familiar tables, such as the 2 or 5 times tables.

Extension: In the second main activity, give these learners the less familiar tables, such as the 3 times table.

Array cards

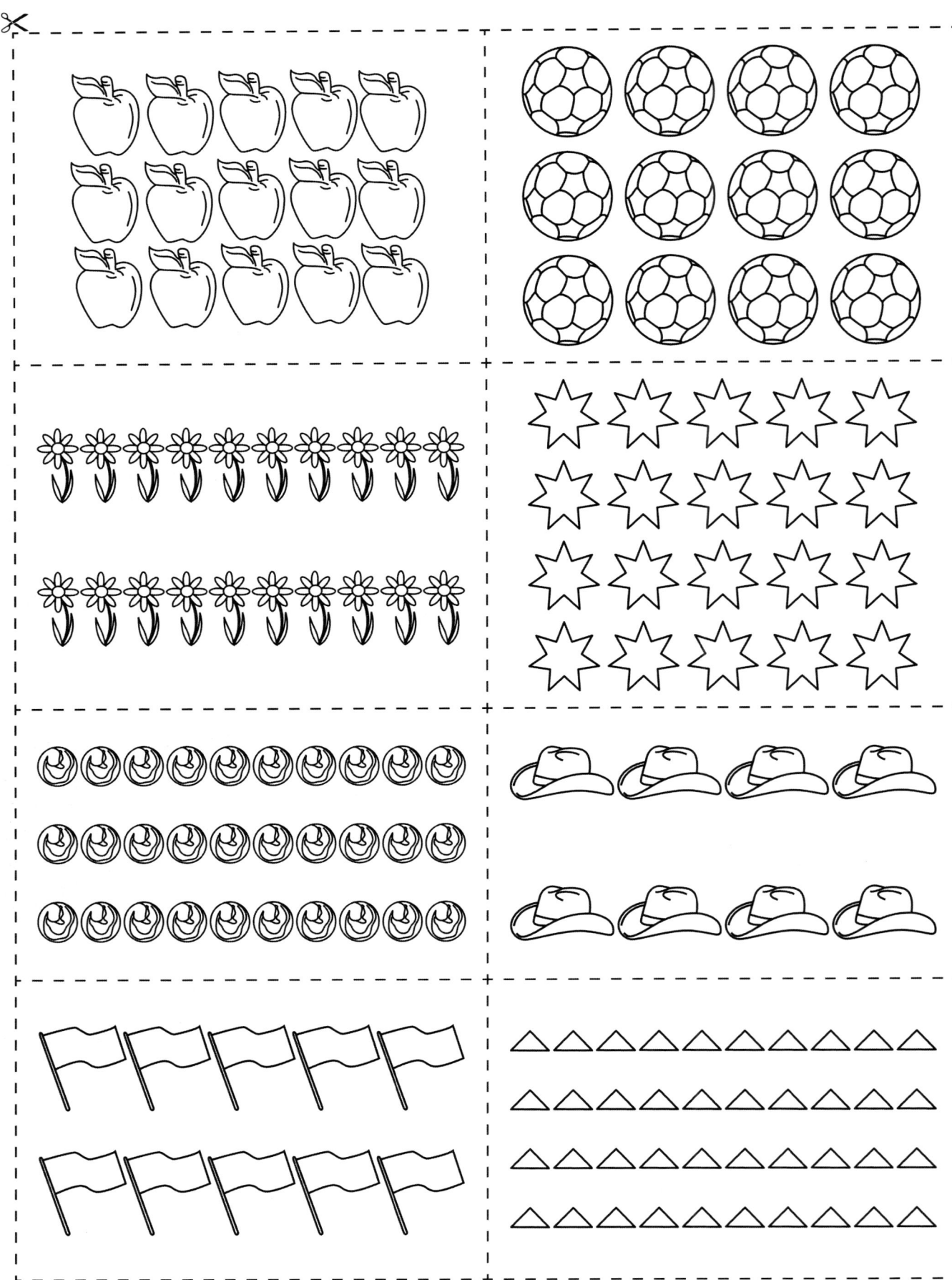

Cambridge Primary: Ready to Go Lessons for Maths Stage 3 © Hodder & Stoughton Ltd 2013

Multiplication and division facts 2

Learning objectives

- Begin to know 4× table. (3Nc4)
- Identify simple relationships between numbers. (3Ps6)

Resources

2 times table chart; 4 times table chart; pencils; plain paper; cards made from photocopiable page 88; calculators; sticky notes; interlocking cubes arranged into lengths of four cubes.

Starter

- Ask quick-fire doubling and halving questions, for example: *Sarah, what's double 9? Ana, what's half of 14?* Adjust questions according to ability, asking most learners to double and halve within 20, the less-able learners within 10 and the more-able learners within 30.

Main activities

- Display the 2 times and 4 times table charts side by side. Highlight the products in both charts. Ask: *What do you notice about these two sets of numbers?* (The numbers in the 4 times table are the numbers in the 2 times table doubled.)
- Take down the times table charts. Demonstrate how to derive facts for the 4 times table by doubling facts from the 2 times table. On the board write: $6 \times 2 = \square$ so $6 \times 4 = \square$. Hand out paper and pencils and ask the learners to copy and complete these multiplication facts. Repeat for other multiples of 2 and 4.
- Demonstrate that dividing by 4 is equivalent to halving and then halving again. Ask the learners to divide multiples of 4 by 4 by halving and halving again.
- Enlarge photocopiable page 88 onto A3 card and cut out the times table cards. Use these to play a game in groups of four: Ask the players to shuffle the cards and place them in a face-down pile, then take it in turns to turn over a card. The first player to say the correct answer keeps the card. The winner is the player with the most cards when all the cards have been turned over. A calculator may be used to check answers.

Plenary

- Display the 4 times table chart. Chant the 4 times table together.
- Ask a volunteer to cover up one of the products on the chart (for example with a sticky note).
- Repeat the previous two steps until all the products are covered up.

Success criteria

Ask the learners:

- Can you write down three facts from the 4 times table?
- Explain how knowing facts in the 2 times table can help you work out facts in the 4 times table.
- What's $28 \div 4$? Explain how you worked out the answer.

Ideas for differentiation

Support: In the game, group these learners together. Remove the divisions from their sets of cards. Give them apparatus (for example interlocking cubes pre-arranged into lengths of four).

Extension: After the game, challenge these learners to derive the facts in the 8 times table from the 4 times table.

4 times table cards

0×4	1×4	2×4
3×4	4×4	5×4
6×4	7×4	8×4
9×4	10×4	$4 \div 4$
$8 \div 4$	$12 \div 4$	$16 \div 4$
$20 \div 4$	$24 \div 4$	$28 \div 4$
$32 \div 4$	$36 \div 4$	$40 \div 4$

Doubling and halving 2

Starter

- Photocopy page 32 onto card and cut out the cards. Remove the cards that involve doubling numbers greater than 10 and halving numbers greater than 20.
- Shuffle the rest of the cards thoroughly and hold them up one at a time.
- Ask the learners to call out the answer. Keep the pace brisk.

Main activities

- Draw the following table on the board.

Starting number	Doubling fact	Halving fact

- Demonstrate how to work out doubles of numbers between 11 and 20 (and the corresponding halves), using known facts and place value knowledge (for example double 17 is double 10 + double 7, double 10 is 20, double 7 is 14; 20 + 14 = 20 + 10 + 4 = 34; so double 17 = 34, and half of 34 = 17). Record the starting number and the doubling and halving facts in the table.

- Ask the learners to choose another number between 11 and 20 and work out the corresponding doubling and halving facts. Record the learners' findings systematically in the table.

- Divide the learners into three ability groups. Ask the learners to copy the table from the board into their books. Ask those in the bottom group to record doubling and halving facts for multiples of 5 between 5 and 30. Ask the middle group to cover multiples of 5 between 35 and 65, and the top group to cover multiples of 5 from 70 to 100.

Plenary

- Display a copy of photocopiable page 90. Ask the learners to provide the numbers for the 'Double' column.

- Ask the learners to describe any patterns they can see in the numbers. Ask whether there are any patterns that they could use to help them remember the doubling facts.

Doubling multiples of 5

Multiple of 5	Double
5	
10	
15	
20	
25	
30	
35	
40	
45	
50	
55	
60	
65	
70	
75	
80	
85	
90	
95	
100	

Cambridge Primary: Ready to Go Lessons for Maths Stage 3 © Hodder & Stoughton Ltd 2013

Multiplication and division strategies 1

Learning objectives

- Multiply single-digit numbers and divide two-digit numbers by 2, 3, 4, 5, 6, 9 and 10. (3Nc21)
- Choose appropriate mental strategies to carry out calculations. (3Pt1)

Resources

Large charts of the 3 and 4 times tables; sticky notes; pencils; plain paper; photocopiable page 92.

Starter

- Challenge the learners to chant the 2, 5 and 10 times tables without looking at times tables charts.
- Display the 3 and 4 times tables charts, and chant them through a few times. Ask volunteers to gradually cover the products (answers) with sticky notes, until all of the products in each table are hidden. Chant through the 3 and 4 times tables once more with all the products hidden.

Main activities

- Give each learner several sheets of paper and a pencil. Ask them to multiply single-digit numbers by 2, 3, 4, 5 and 10. Discuss strategies used (for example 'just knowing' the multiplication fact, counting on in steps either mentally or on a sketched number line, drawing an array, and so on).
- Ask the learners to multiply single-digit numbers by 6 and 9. Discuss strategies used (for example to multiply by 9, multiply by 10 and compensate with subtraction; to multiply by 6, multiply by 5 and compensate with addition, or multiply by 3 and double the answer).

- Display a multiplication grid from photocopiable page 92, and explain how the grid works. Demonstrate how to use it to find answers to multiplication and division questions.
- Give each learner a multiplication grid. In pairs, ask the learners to take turns to ask and answer multiplication questions. The learner asking the question should look at their grid and the learner answering the question should put their grid out of sight.

Plenary

- Give the learners two-digit numbers to divide by 2, 3, 4, 5, 6, 9 and 10 (no remainders). Discuss strategies used.
- Display a multiplication grid and highlight the 9 times table. Ask the learners to describe patterns in it.

Success criteria

Ask the learners:

- What is 9×3?
- Divide 35 by 5.
- Multiply 8 by 6. Explain how you worked out the answer.

Ideas for differentiation

Support: In the final main activity, give these learners a copy of the multiplication grid on which you have folded over or cut off the bottom five rows.

Extension: In the final main activity, ask these learners to include division questions as well as multiplication questions.

Multiplication grid

×	0	1	2	3	4	5	6	7	8	9	10
0	0	0	0	0	0	0	0	0	0	0	0
1	0	1	2	3	4	5	6	7	8	9	10
2	0	2	4	6	8	10	12	14	16	18	20
3	0	3	6	9	12	15	18	21	24	27	30
4	0	4	8	12	16	20	24	28	32	36	40
5	0	5	10	15	20	25	30	35	40	45	50
6	0	6	12	18	24	30	36	42	48	54	60
7	0	7	14	21	28	35	42	49	56	63	70
8	0	8	16	24	32	40	48	56	64	72	80
9	0	9	18	27	36	45	54	63	72	81	90
10	0	10	20	30	40	50	60	70	80	90	100

×	0	1	2	3	4	5	6	7	8	9	10
0	0	0	0	0	0	0	0	0	0	0	0
1	0	1	2	3	4	5	6	7	8	9	10
2	0	2	4	6	8	10	12	14	16	18	20
3	0	3	6	9	12	15	18	21	24	27	30
4	0	4	8	12	16	20	24	28	32	36	40
5	0	5	10	15	20	25	30	35	40	45	50
6	0	6	12	18	24	30	36	42	48	54	60
7	0	7	14	21	28	35	42	49	56	63	70
8	0	8	16	24	32	40	48	56	64	72	80
9	0	9	18	27	36	45	54	63	72	81	90
10	0	10	20	30	40	50	60	70	80	90	100

Multiplication and division strategies 2

Learning objectives

- Understand that division can leave a remainder (initially as 'some left over'). (3Nc24)
- Make sense of and solve word problems, single (all four operations) and two-step (addition and subtraction), and begin to represent them, e.g. with drawings or on a number line. (3Pt3)
- Check a division using multiplication, e.g. check 12 ÷ 4 = 3 by doing 4 × 3. (3Pt7)

Resources

Pencils; plain paper; photocopiable page 94.

Starter

- Hand out several sheets of paper and pencils.
- Demonstrate dividing two-digit numbers by single-digit numbers (no remainder) using grouping (for example calculate 35 ÷ 5 by drawing groups of five dots until you've drawn 35, and then counting the number of groups). Show how to check answers using multiplication (for example if 35 ÷ 5 = 7, then 7 × 5 should equal 35).
- Give the learners similar divisons to do.
- Demonstrate how to record the answer when a division has 'some left over' (for example 22 ÷ 4). Introduce the term 'remainder' and the notation used. Show how to check answers using multiplication and addition (for example if 22 ÷ 4 = 5 r2, then (5 × 4) + 2 should equal 22).
- Give the learners similar divisions to do.

Main activities

- Display a copy of photocopiable page 94. Read one of the problems aloud. Ask: *What calculation do you need to do in order to solve this problem?* Write the calculation on the board.

- Ask the learners to estimate the answer and explain their reasoning.
- Give them a few minutes to perform the calculation (either individually or in pairs). Ask: *What answer did you get? Is your answer reasonable? How do you know?*
- Ask volunteers who got the correct answer to explain their calculation strategies.
- Repeat the process for a second problem.
- Give out photocopiable page 94. Ask the learners to work through the rest of the problems independently or in pairs.

Plenary

- Read through a couple of the problems that you didn't work through together during the main activity. Ask the learners to share their answers, and explain their methods.

Success criteria

Ask the learners:

- What calculation did you need to do in order to solve this problem?
- What is the answer to the problem? How did you work it out?
- How could you check your answer?

Ideas for differentiation

Support: The problems on photocopiable page 94 are arranged from easiest to hardest; it may be appropriate for these learners to attempt only the first five questions.

Extension: These learners should work individually rather than with a partner. Early finishers could devise similar problems to give to a friend.

Name: _____

Multiplication and division problems

1. Eliana is organising her socks.
   She makes nine pairs and has one sock left over.
   How many socks does she have altogether?

   _____

2. Sam draws a regular pentagon.
   The perimeter of the pentagon is 40 cm.
   How long is each side?

   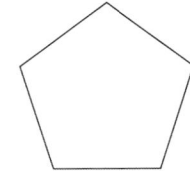

   _____

3. There are 34 children in Class 3.
   The teacher asks them to get into groups of four.
   How many groups of four will there be, and how many children
   will be left over?

   _____

4. At a party some children ate 16 pizza slices.
   Each slice is one quarter of a whole pizza.
   How many whole pizzas did they eat?

   _____

5. 51 children are going on a school trip. There are three buses.
   The children are divided equally among the buses.
   How many children are on each bus?

   _____

6. Pilar buys 12 CDs for $9 each. How much does she spend altogether?

   _____

7. Arjun has 45 eggs and some egg boxes. He must put each egg in an egg box.
   Each egg box can hold six eggs. How many egg boxes does Arjun need?

   _____

Cambridge Primary: Ready to Go Lessons for Maths Stage 3 © Hodder & Stoughton Ltd 2013

Unit assessment

- If you partition 784 you get 700 + 80 + 4. What do you get if you partition: a) 392 b) 860 c) 509?
- Tell me a multiplication fact that you know. What division fact follows from that multiplication fact?

- Write a three-digit number. Which number is: a) 1 less b) 10 more c) 100 less?
- Can you write a list, draw a table or draw a diagram to show pairs of multiples of 5 that total 100 (for example 25 + 75 = 100)?

Summative assessment activities

Observe the learners while they play these games. You will quickly be able to identify those who appear to be confident and those who may need additional support.

Ordering numbers game

This game assesses the learners' ability to order two- and three-digit numbers.

You will need:

Blank cards; pencils; timer.

What to do

- Organise the learners into groups. Give each learner 12 blank cards and a pencil. Ask the learners to write a different two- or three-digit number on each card, and place the cards in a single pile, face down.
- One player should act as timekeeper. The timekeeper must shuffle the cards and put a row of six cards face down. The active player (to the timekeeper's left) must turn over these cards and order them from largest to smallest, while the timekeeper times how long it takes. The active player scores 3 points for a time less than 5 seconds, 2 points for between 5 and 10 seconds and 1 point for between 10 and 20 seconds. The active player becomes the timekeeper in the next round. The winner is the player with the most points when all the cards have been used.

Doubling and halving game

This game assesses the learners' ability to work out quickly the doubles of numbers 1 to 20 and the related halves.

You will need:

Soft ball.

What to do

- Ask the learners to stand in a circle. Throw the soft ball to the learner on your left, calling out a doubling question (1 to 20), for example: *Double fourteen!* The learner must give the answer before calling out their own doubling question, and throwing the ball to the learner on their left.
- Play a second round of the game in which each question is a halving question (even numbers to 40).

Distribute photocopiable page 96. Ask the learners to read and answer the questions. They should work independently.

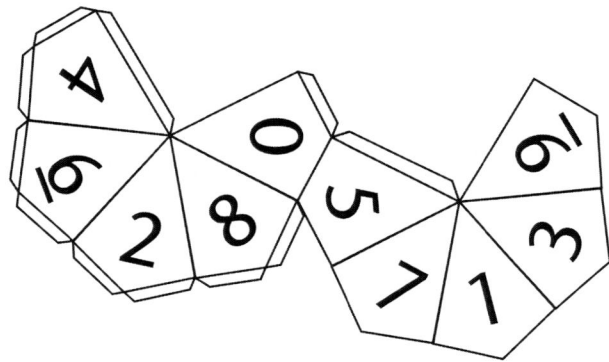

Name: _____

Looking at numbers

1. Write < or > between each pair of numbers.

   a) 936 ☐ 639

   b) 210 ☐ 201

   c) 342 ☐ 423

   d) 181 ☐ 118

2. Draw a number line showing the position of each of these numbers:

   496, 511, 483, 529, 504, 542, 535.

3. Round these numbers to the nearest 10:

   a) 68 _____

   b) 45 _____

   c) 81 _____

   d) 54 _____

4. Round these numbers to the nearest 100:

   a) 134 _____

   b) 432 _____

   c) 750 _____

   d) 967 _____

5. 235 + 67 _____

6. 506 – 9 _____

7. Write the missing numbers.

   a) ☐ × 4 = 32

   b) 6 × ☐ = 18

   c) 7 × 5 = ☐

8. There are 33 children in Class 4.
   The teacher asks them to get into groups of five.
   How many groups of five will there be, and how many children
   will be left over?

_____

Money 3

- Consolidate using money notation. (3Mm1)
- Use addition and subtraction facts with a total of 100 to find change. (3Mm2)
- Make a sensible estimate for the answer to a calculation, e.g. using rounding. (3Pt11)

Pencils; plain paper; cards made from photocopiable page 98; lots of price-labelled items (all prices less than $4, most prices multiples of 5c); US coins; $1 bills; $5 bills.

Starter

- Hand out paper and pencils to the learners.
- Enlarge photocopiable page 98 onto A3 card to make large money cards. Display three of the money cards and write the amount of money shown on one of the cards (for example $2.45). Read the amount of money aloud. Ask the learners to indicate which of the three cards shows that amount of money by writing down its letter. Repeat several times.
- Display a single money card. Ask the learners to write down the amount of money on the card, using correct money notation. Repeat several times.

Main activities

- Model finding change from whole dollar amounts (prices multiples of 5c only). Begin with tendering $1 and finding change for prices between 5c and 95c, for example find the change from $1 for a spend of 45c. Extend to finding change from $2, $3 or $4, for example find the change from $3 for a spend of $2.15.
- Discuss the practice of making an estimate of the amount of change you will get before a transaction, and then checking the amount you receive against that estimate, to make sure you have been given the correct amount.

- Model recording transactions, for example price $2.65; money paid $3; change given 35c.
- Distribute the price-labelled items, coins and $1 bills. Ask the learners to work in pairs to sell each other items. The buyer should offer the seller an appropriate whole number of dollars. The seller makes and gives the correct change. The buyer uses estimation to check that they have got the correct change. Both learners then record the transaction.

Plenary

- Write a price on the board that is less than $4, but is not a multiple of 5c (for example $3.27). Ask: *How much change would you get from $4?* Allow the learners to work out the answer with a partner, using coins.
- Discuss the strategies used.

Ask the learners:

- (Pointing to a collection of coins:) How much money is this? Write down the amount.
- If an item cost $1.25 and you paid for it with $2, how much change would you get?
- Explain how you worked out the answer.

Support: Give these learners items priced under $2 in multiples of 5c.

Extension: Give these learners $5 bills as well as $1 bills, so that they get the opportunity to calculate amounts of change greater than $1.

Money cards

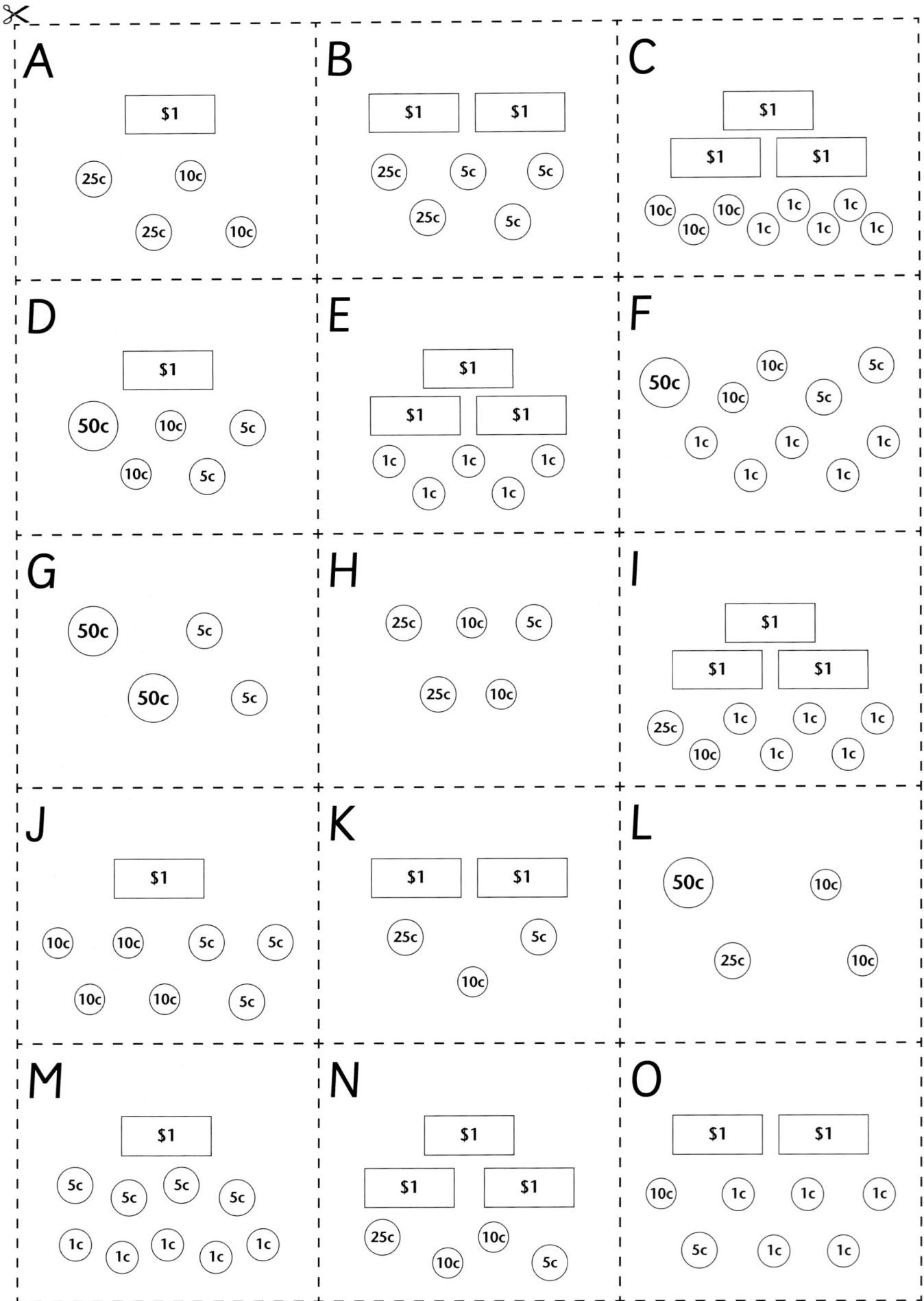

A

| $1 |

25c   10c

25c   10c

B

| $1 | | $1 |

25c   5c   5c

25c   5c

C

| $1 |

| $1 | | $1 |

10c 10c 10c 1c 1c

10c 1c 1c 1c

D

| $1 |

50c 10c 5c

10c 5c

E

| $1 |

| $1 | | $1 |

1c 1c 1c

1c 1c

F

50c 10c 5c

10c 5c

1c 1c 1c

1c 1c

G

50c 5c

50c 5c

H

25c 10c 5c

25c 10c

I

| $1 |

| $1 | | $1 |

25c 1c 1c 1c

10c 1c 1c

J

| $1 |

10c 10c 5c 5c

10c 10c 5c

K

| $1 | | $1 |

25c   5c

10c

L

50c   10c

25c   10c

M

| $1 |

5c 5c 5c 5c

1c 1c 1c 1c 1c

N

| $1 |

| $1 | | $1 |

25c 10c 5c

10c

O

| $1 | | $1 |

10c 1c 1c 1c

5c 1c 1c

Money 4

- Solve word problems involving measures. (3MI5)
- Choose appropriate mental strategies to carry out calculations. (3Pt1)
- Consider whether an answer is reasonable. (3Pt12)
- Make up a number story to go with a calculation, including in the context of money. (3Ps1)

Pencils; plain paper; photocopiable page 100; US coins; $1 bills; $5 bills; price cards with prices less than $5.

Starter

- Model finding the complement to 100 for any whole number. Sketch a number line on the board, placing the starting number on it and counting up in ones to the next multiple of 10, and then in 10s to 100.
- Hand out paper and pencils. On the board write a whole number less than 100. Ask the learners to write down the complement to 100 using the number line method you demonstrated. Repeat for at least half a dozen numbers.
- Talk about when this method might be useful in situations handling money (for example when calculating change from a whole number of dollars).

Main activities

- Display question 1 from photocopiable page 100. Read through it together and then ask: *What calculation do you need to do in order to solve this problem? How could you perform the calculation?* Repeat for questions 2, 3 and 4.
- Display question 5 from photocopiable page 100. Ask the learners to work in pairs to write a word problem to go with this calculation. Ask volunteers to read out their word problems. Repeat for question 6.

- Organise the learners into pairs. Give each pair access to coins, bills and price cards. Ask each pair to use price cards to write four money calculations. Ask them to then write a word problem to go with each calculation. Finally, ask them to swap problems with another pair and solve each other's problems.

Plenary

- Ask individual learners to give their answers to questions 1 to 4 on photocopiable page 100. Ask the other learners to say whether they think the given answer is reasonable, and explain their reasoning.
- Ask several pairs to share word problems they've written. Ask the rest of the class to solve the problems, and explain how they worked out the answer.

Ask the learners:

- Can you read me a word problem that you've written?
- What is the calculation your word problem is based on?
- What is the answer to that calculation?
- How did you work it out?

Support: Group these learners together and give them price cards with prices that are multiples of 5c only. Work with them to devise their word problems.

Extension: Challenge these learners to write one word problem for each type of calculation (addition, subtraction, multiplication and division).

Name: _____

Money 4

1. Lila buys an ice-cream for $1.80 and a fruit juice for $1.45.
   How much does she spend altogether?

2. Jin buys a model car for $2.85. He pays with a $5 bill.
   How much change should he get?

3. Every school day Laura spends 60c on the fare for the school bus.
   Laura goes to school five days a week.
   How much does she spend on school bus fares every week?

4. Nabarun spends $3.15 buying three identical pens.
   How much does each pen cost?

5. $3 – $2.43

6. $1.15 × 4

Cambridge Primary: Ready to Go Lessons for Maths Stage 3 © Hodder & Stoughton Ltd 2013

Length 2

Learning objectives

- Read to the nearest division or half division, use scales that are numbered or partially numbered. (3Ml3)
- Use a ruler to draw and measure lines to the nearest centimetre. (3Ml4)
- Explain a choice of calculation strategy and how the answer was worked out. (3Ps2)
- Use ordered lists and tables to help solve problems systematically. (3Ps4)

Resources

Number line made from photocopiable page 102; tape measures; rulers.

Starter

- Enlarge photocopiable page 102 onto A3 card and make the number line. Display the number line and ask the learners how many equal parts each whole number is divided into. (Ten.) Establish that each of these parts is one-tenth.
- Practise counting on and back in tenths from any whole number (for example three, three and one-tenth, three and two-tenths, three and three-tenths, ...).
- Identify halves on the number line. Establish that they are the numbers halfway between the whole numbers, and that one half is equivalent to five-tenths.
- Point to an unlabelled division. Ask the learners to say the number aloud (for example two and seven tenths), and then round it to the nearest half (two and a half).

Main activities

- Demonstrate how to use a tape measure to measure lengths to the nearest half centimetre. Link to the work done in the starter activity.

- Organise the learners into groups of four or six and give each pair within the group a tape measure. Write a list of body measurements on the board (for example height, arm length, leg length, wrist circumference, head circumference). Ask pairs to take each other's measurements to the nearest half centimetre and record all the group's measurements in a table.
- Ask the learners to make an ordered list for each body measurement within their group that lists the group members' names in order, together with the measurement, from smallest to largest measurement.
- Write two body measurements on the board, asking the learners to find the difference between them. Ask them how they worked out the answer, and why they chose the method they used.

Plenary

- Ask the learners questions that require them to analyse and interpret the data they have collected in the table and ordered lists, for example: *Do people with longer arms also have longer legs? Are the people with the longest legs always the tallest?*

Success criteria

Ask the learners:

- What is your hand span to the nearest half centimetre?
- Draw a line that is $12\frac{1}{2}$ centimetres long.
- Who is the second tallest person in your group?
- Who has the second smallest wrist measurement in your group?

Ideas for differentiation

Support: Group these learners together. Check that they are taking and rounding measurements correctly and help them to order measurements.

Extension: Ask these learners to calculate using measured lengths, for example: *Find the difference between your height and your friend's height.*

0 to 5 number line marked in tenths

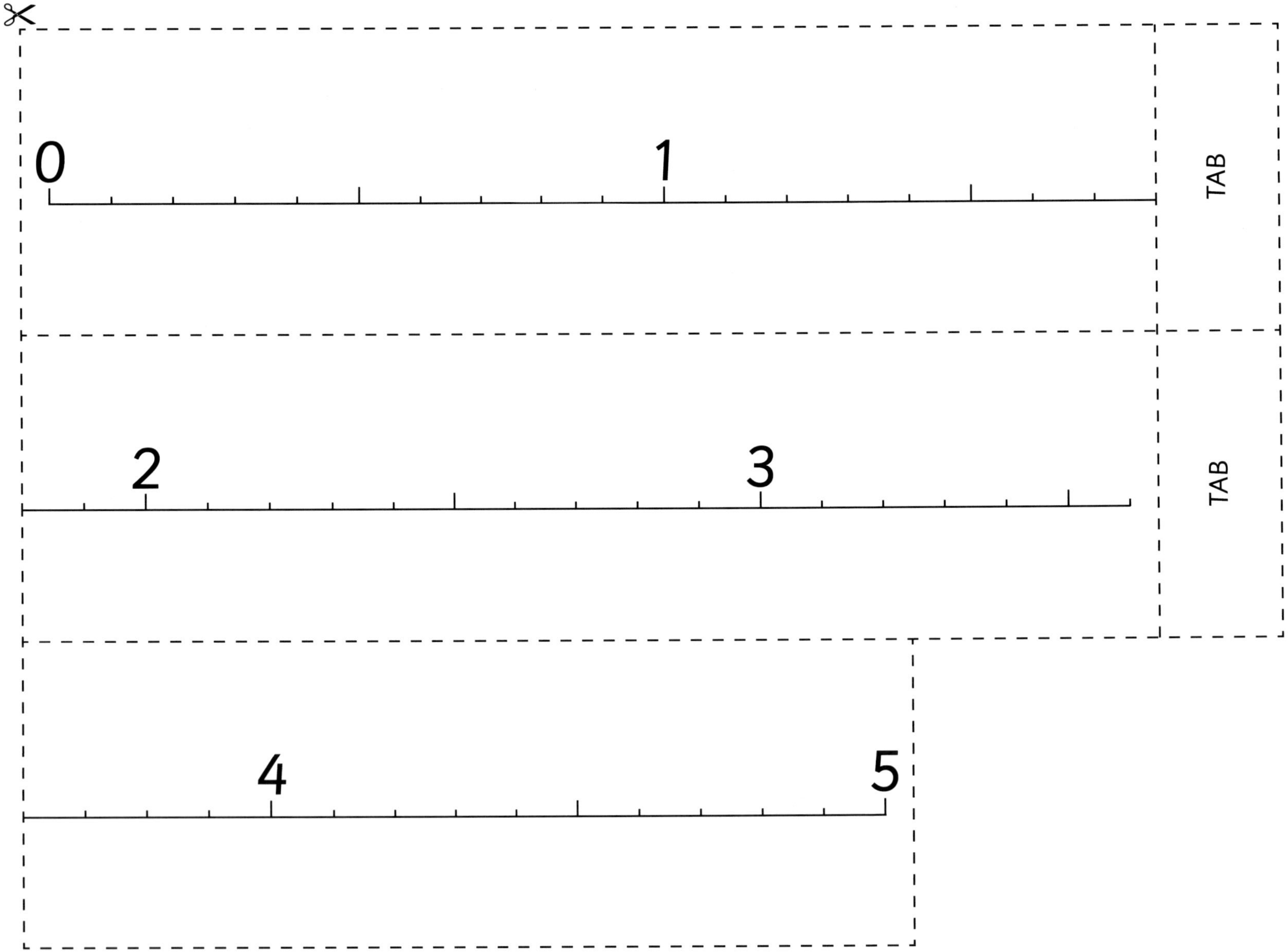

0

1

2

3

4

5

TAB

TAB

Cambridge Primary: Ready to Go Lessons for Maths Stage 3 © Hodder & Stoughton Ltd 2013

Time 4

Learning objectives

- Solve word problems involving measures. (3Ml5)
- Suggest and use suitable units to measure time and know the relationships between them (second, minute, day, hour, week, month, year). (3Mt1)
- Read a calendar and calculate time intervals in weeks or days. (3Mt4)

Resources

Photocopiable page 104; timer, one display copy and plenty of tabletop copies of a calendar (the current month plus the next one or two months).

Starter

- Revise relationships among the various units of time by playing a follow-me card game:
  - Enlarge photocopiable page 104 onto A3 card and cut out the time cards. Find and keep the START card. Hand out the rest of the cards. Some learners may need to share one card between two.
  - Read the START card aloud. The learner who has the answer to that question written on their card must read their card aloud, and so on, until the END card is reached.
  - Play a few more rounds, redistributing the cards between rounds. Time the activity, and challenge the class to beat their previous best time.

Main activities

- Display a large copy of a calendar for the current month, and the following one or two months. Before the lesson, mark on the calendar a variety of real or imaginary events (for example the learners' birthdays and school and community events).
- Ask the learners questions about the events on the calendar that require them to work out time intervals in weeks or days, for example: *How long is it until the end of term? How far apart are Carlos's and Ranjit's birthdays?* Discuss different ways of expressing the same duration of time (for example 23 days or 3 weeks 2 days).

- Organise the learners into pairs and give each pair a copy of the calendar. Write a few more calendar problems on the board, asking the learners to work in pairs to solve them, and write down their answers.
- Ask the learners to work in pairs to write similar calendar problems and swap problems with another pair.

Plenary

- Ask volunteers to share the calendar-related word problems they have written. Ask the rest of the class to solve the problems. Discuss answers and strategies used.

Success criteria

Ask the learners:

- How long ago was … [past event]?
- How long will you have to wait until … [future event]?
- What's another way of saying '17 days'?
- Read me a calendar problem you have written. What is the answer?

Ideas for differentiation

Support: Group these learners together and support them when they're working on the problems written on the board. Ask each pair to write only one or two problems of their own.

Extension: Challenge these learners to write a greater number of word problems for another pair to solve.

Time follow-me cards

START How many hours in a day?	**24** How many minutes in a quarter of an hour?	**15** How many seconds in half a minute?	**30** How many minutes in an hour?
60 How many days in a leap year?	**366** How many months in a year?	**12** How many hours in two days?	**48** How many days in a week?
7 How many days in February in a leap year?	**29** How many weeks in 14 days?	**2** What fraction of a year is six months?	**Half** Roughly how many weeks in a month?
4 How many years is 36 months?	**3** How many seconds in 2 minutes?	**120** How many days in a non-leap year?	**365** How many days in February in a non-leap year?
28 How many weeks in a year?	**52** How many days in January?	**31** How many days in 3 weeks?	**21** How many days from Saturday to Thursday?
5 How many months from February to October?	**8** How many months in half a year?	**6** How many hours between midday and 9 o'clock at night?	**9** END

Time 5

- Read the time on analogue and digital clocks, to the nearest 5 minutes on an analogue clock and to the nearest minute on a digital clock. (3Mt2)

One large and plenty of small cardboard clock faces with moveable hands; pencils; plain paper; photocopiable page 106; plain card; scissors.

Starter

- Give each learner a small clock face. Call out a time (intervals of 5 minutes only), for example: *Twenty past eight, A quarter to five, Nine fifty-five.* Ask the learners to make the time you call out on their clock face. Confirm the correct answer by displaying the time on the large clock face. Repeat for a variety of times.

- Give each learner some paper and a pencil. Make a time on the large clock face and ask the learners to write it down. Compare alternative forms of correct answers (for example 15 minutes past 2, 2.15, a quarter past 2).

Main activities

- Introduce the concepts of a.m. and p.m. Establish that both midday and midnight are 12.00. Use a diagram like the one below.

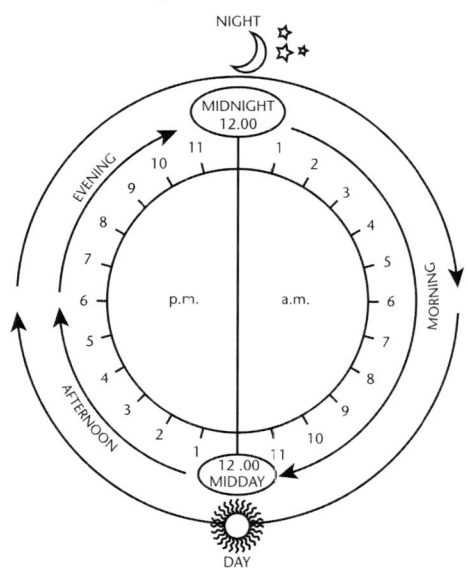

- Using questions, give the learners practice in expressing times of day using a.m. and p.m. You could base your questions around times in the school day, for example: *What time does school start? What time do we have morning break / lunch / afternoon break? What time does school finish?*

- Enlarge photocopiable page 106 onto A3 card to make time match cards. Organise the learners into pairs and give each pair a set of time match cards. Ask pairs to sort the cards into groups, each group of cards showing the same time written in different ways.

- Ask pairs of learners to make their own sets of time match cards. Pairs from a single table should pool their cards then swap cards with other tables and try to match them up.

Plenary

- Display about half a dozen time cards in a random order. The times may be expressed in various ways, but each time should be different. Ask the learners to order the times from earliest to latest.

Ask the learners:

- What time does this analogue clock show?
- What time does this digital clock show?
- Write a quarter to eight on a digital clock.
- Make four fifty-five on an analogue clock.

Support: In the starter activity, pair these learners with learners who are more confident in telling the time.

Extension: Challenge these learners to create time match cards that show the time in one-minute intervals.

Time match cards

3.45a.m.	a.m.	quarter to four in the morning	45 minutes past 3a.m.
3.45p.m.	p.m.	quarter to four in the afternoon	45 minutes past 3p.m.
11.40a.m.	a.m.	twenty minutes to midday	40 minutes past 11a.m.
11.40p.m.	p.m.	twenty minutes to midnight	40 minutes past 11p.m.
8.50p.m.	p.m.	ten to nine in the evening	50 minutes past 8p.m.
8.50a.m.	a.m.	ten to nine in the morning	50 minutes past 8a.m.

Time 6

Learning objectives

- Solve word problems involving measures. (3Ml5)
- Begin to calculate simple time intervals in hours and minutes. (3Mt3)
- Estimate and approximate when calculating, and check working. (3Pt10)

Resources

Counting stick; two large and plenty of small clock faces with moveable hands; photocopiable page 108.

Starter

- To use the counting stick, hold it horizontally, and point to the left end as viewed by the learners (your right). Point to each division in turn, counting aloud as you go. To count down, start at the right end of the stick (your left).
  - Count on in whole hours from any time (for example seventeen minutes past ten, seventeen minutes past eleven, seventeen minutes past twelve, seventeen minutes past one …).
  - Count on in intervals of five minutes from times given to the nearest five minutes (for example five fifty, five fifty-five, six, six-oh-five …).
  - Count on in intervals of one minute from any time (for example eight thirty-eight, eight thirty-nine, eight forty, eight forty-one …).

Main activities

- Use the large clock faces to display two times over 1 hour apart. Demonstrate finding the interval between them by counting on in intervals of 1 hour, and then 5 minutes. Repeat for other pairs of analogue times.
- On the board write two digital times over 1 hour apart (not multiples of 5 minutes). Demonstrate finding the interval between the two times by counting on in intervals of 1 hour, then 5 minutes, and finally 1 minute. Repeat for other pairs of digital times.

- Display a copy of photocopiable page 108. Read through the first problem.
- Ask the learners to estimate the answer, for example 'some time between one o'clock and half past one'. Record an estimate.
- Ask the learners to solve the problem (in pairs or individually) and check their answers by comparing them to the estimate.
- Repeat the process for a second problem.
- Organise the learners to work in pairs, giving each pair photocopiable page 108 and two small clock faces.

Plenary

- Read through a few of the word problems that you have not already worked through together. Ask the learners to explain how they worked out the answers. Ask whether anyone else used a different method to solve the same problem.

Success criteria

Ask the learners:

- What did you need to do to solve this problem?
- What was your estimate? Explain how you worked it out.
- What answer did you get? Explain how you worked it out.
- Is your answer reasonable? How do you know?

Ideas for differentiation

Support: Group these learners together and guide them through an extra problem. Ask them to complete only the first four problems, as these are the easiest.

Extension: Ask these learners to make up their own time problems and give them to a friend to solve.

Name: _____

Time problems 2

1. Omar puts a cake in the oven at 11.25a.m.
    It needs cooking for 1 hour 50 minutes.
    At what time will Omar need to take the cake out of the oven?

    _____

2. Juan rents a movie that lasts 1 hour and 35 minutes.
    He starts watching the movie at 7.45p.m.
    What time will the movie end?

    _____

3. A concert begins at 6.45p.m. and ends at 8.30p.m.
    How long does the concert last?

    _____

4. The Ramirez family have been travelling for 2 hours 10 minutes.
    It is now 1.35p.m. At what time did they start their journey?

    _____

5. An aeroplane was delayed by 3 hours 42 minutes.
    It was supposed to arrive at 10.15a.m.
    What time did it actually arrive?

    _____

6. Ambika wakes up at 7.24a.m. after sleeping for 8 hours 38 minutes.
    What time did she go to sleep?

    _____

7. In a marathon, Hamid crossed the start line at 9.57a.m. and crossed
    the finishing line at 2.04p.m. His friend Cheng crossed the start line at
    10.01a.m. and crossed the finishing line at 1.58p.m.

    a) What was Hamid's time for the race? _____

    b) What was Cheng's time for the race?_____

    c) Who ran the race in the faster time? _____

    d) What is the difference between the two friends' times? _____

Cambridge Primary: Ready to Go Lessons for Maths Stage 3 © Hodder & Stoughton Ltd 2013

Mass 2

Learning objectives

- Choose and use appropriate units and equipment to estimate, measure and record measurements. (3MI1)
- Read to the nearest division or half division, use scales that are numbered or partially numbered. (3MI3)
- Solve word problems involving measures. (3MI5)
- Begin to understand everyday systems of measurement in length, weight, capacity, time and use these to make measurements as appropriate. (3Pt2)

Resources

Pencils; plain paper; cards made from photocopiable page 110; variety of instruments for measuring mass (e.g. pan balances, spring scales, kitchen scales, bathroom scales); a brick.

Starter

- Organise the learners into pairs and give each pair paper and pencils.
- Copy photocopiable page 110 onto card to make mass cards. Hold up the mass cards one at a time. Ask the learners to write the mass shown on each card. Discuss answers, talking about how each scale works, and look at various correct ways of recording the same mass (for example 1 kilogram 200 grams = 1 kg 200 g = 1200 grams = 1200 g). Do not bring up decimal notation (for example 1.2 kg) unless a learner does so.

Main activities

- Revise the names of the different instruments for measuring mass and how to use each one. Discuss the approximate range of mass that each instrument is designed for measuring.
- Choose two objects from the classroom (for example a large hardback book and a small paperback book). Use appropriate instruments to find and record their masses (for example weigh each book on a pan balance and record its mass in grams). Use these measurements to write a word problem involving addition, subtraction or multiplication (for example 'I am going on a day trip and my bag is too heavy. I take out this book and replace it with this book. How much lighter will my bag be?'). Ask the learners to solve the problem, and then discuss calculation strategies. Repeat for a different pair of objects.

- Organise the learners into groups of two or three to choose pairs of objects, measure their masses and use these measurements to write a word problem. Once groups have written four word problems, ask them to swap problems with another group, and solve each other's problems.

Plenary

- Ask volunteers to read out a word problem they've written. Ask the rest of the class to solve the problems, and then explain how they worked them out.
- Discuss methods for checking the reasonableness of answers (for example using rounding and approximation, redoing the calculation using another method, or using the inverse operation: subtraction to check addition and vice versa, or multiplication to check division and vice versa).

Success criteria

Ask the learners:

- Which instrument would you use to measure the mass of a brick? Why?
- Which unit would you use to measure the mass of a brick?
- Measure the mass of a brick.
- Read out a word problem you solved. Explain how you solved it.

Ideas for differentiation

Support: When organising groups for the final main activity, make sure these learners are working with more confident learners who work well with others.

Extension: Challenge these learners to write an answer sheet for the problems they write, and to include at least one division problem.

Mass cards

Cambridge Primary: Ready to Go Lessons for Maths Stage 3 © Hodder & Stoughton Ltd 2013

Capacity 2

- Choose and use appropriate units and equipment to estimate, measure and record measurements. (3MI1)
- Know the relationship between kilometres and metres, metres and centimetres, kilograms and grams, litres and millilitres. (3MI2)
- Read to the nearest division or half division, use scales that are numbered or partially numbered. (3MI3).
- Solve word problems involving measures. (3MI5)

Pencils; plain paper; cloths; cards made from photocopiable page 112; a variety of measuring jugs / cups / bowls marked in millilitres; a variety of waterproof containers (e.g. jugs, bottles, bowls, jars, mugs, cups, spoons, ladles); water.

Starter

- Organise the learners into pairs and hand each pair paper and pencils.
- Enlarge photocopiable page 112 onto A3 card to make volume cards. Hold up the cards one at a time and ask the learners to write the volume shown on each card. Discuss answers, talking about how each scale works, and look at various correct ways of recording the same volume (for example 1 litre 200 millilitres = 1 litre 200 ml = 1200 millilitres = 1200 ml). Do not bring up decimal notation (for example 1.2 litre) unless a learner does so.

Main activities

- Revise how to use measuring jugs, cups or bowls marked in millilitres to measure the capacity of various waterproof containers. Emphasise the importance of placing the measuring container on a flat surface and moving so that your eyes are level with the top of the water.
- Choose two waterproof containers. Use appropriate measuring jugs, cups or bowls to find and record their capacities. Use these measurements to write a word

problem involving addition, subtraction or multiplication, for example: *You fill this cup with water and pour the water into this jug. How much more water could you pour into the jug?* Ask the learners to solve the problem, and then discuss calculation strategies. Repeat for a different pair of containers.

- Organise the learners into groups of two or three to choose pairs of containers, measure their capacities and use these measurements to write a word problem. Once groups have written four word problems, ask them to swap problems with another group, and solve each other's problems.

Plenary

- Ask the learners to describe any problems they encountered when measuring the capacity of the containers, and how they solved them (for example using too large a measuring container so that the water poured into it does not reach the bottom of the scale).
- Ask volunteers to read out a word problem they've written. Ask the rest of the class to solve the problems, and then explain how they worked them out.

Ask the learners:

- Which measuring container would you use to measure the capacity of a ladle? Why?
- Which unit would you use to measure the capacity of a ladle – millilitres or litres?
- Measure the capacity of a ladle.
- Read out a word problem you solved. Explain how you solved it.

Support: When organising groups for the final main activity, make sure the less-able learners are working with more confident learners who work well with others.

Extension: Challenge the more-able learners to write an answer sheet for the problems they write, and to include at least one division problem.

Volume cards

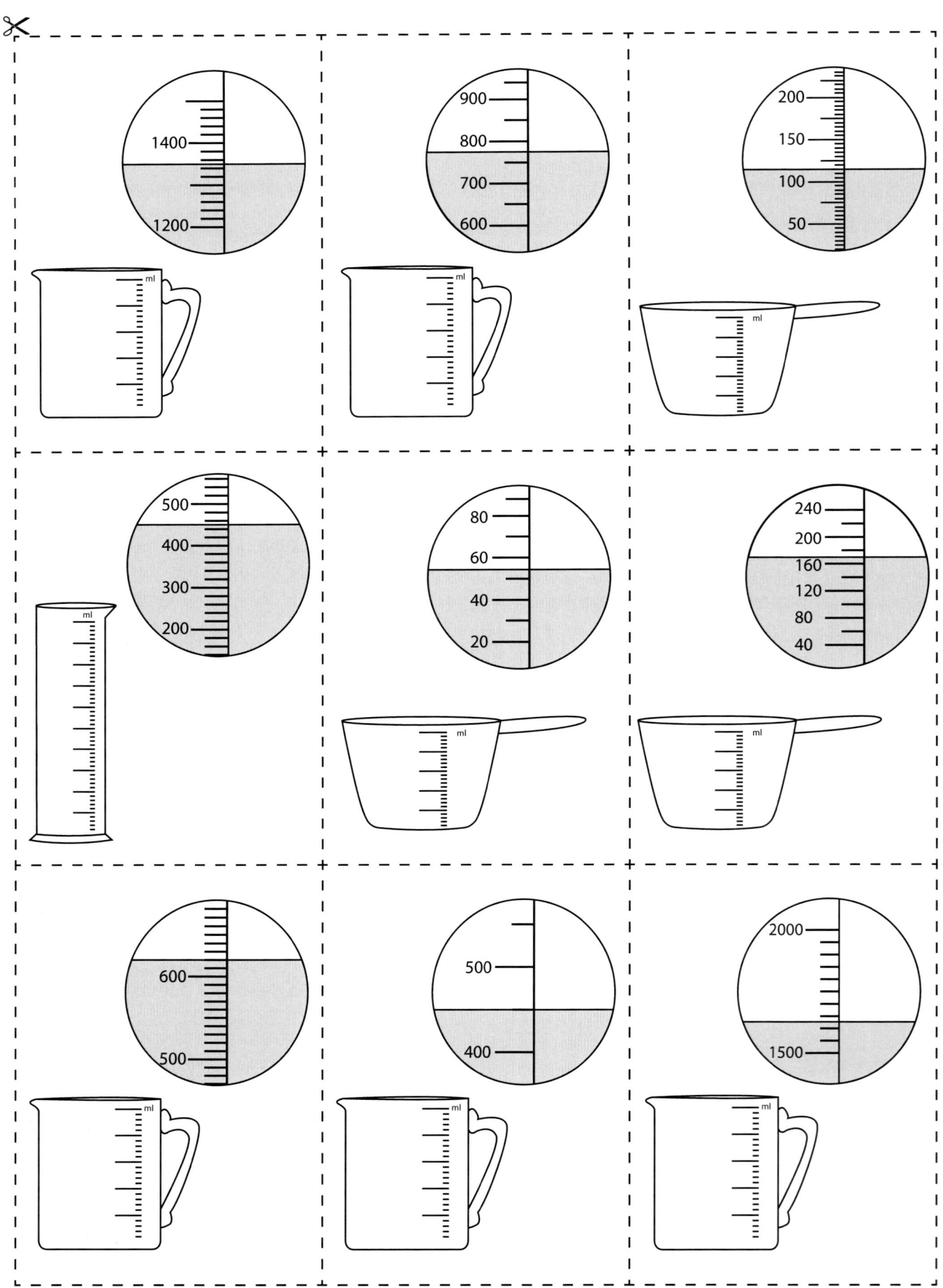

Cambridge Primary: Ready to Go Lessons for Maths Stage 3 © Hodder & Stoughton Ltd 2013

Unit assessment

Questions to ask

- Can you name two units that are used to measure mass? What is the relationship between them?
- Can you estimate the mass of an apple? How did you make your estimate?
- What units would you use to measure: a) the capacity of a bath tub and b) the capacity of a spoon?
- Can you write four dollars and five cents in figures using the dollar sign?

Summative assessment activities

Observe the learners while they take part in these activities. You will quickly be able to identify those who appear to be confident and those who may need additional support.

Telling the time

This activity assesses the learners' ability to tell the time to the nearest five minutes on an analogue clock.

You will need:

A large clock face with moveable hands; small clock faces with moveable hands; pencils; plain paper.

What to do

- Give each learner a small clock face. Call out a time (intervals of 5 minutes only), for example: *Ten minutes to nine, One thirty-five.* Ask the learners to make the time on their clock face. Confirm the correct answer by displaying the time on the large clock face. Repeat for various times.
- Give each learner paper and a pencil. Make a time on the large clock face and ask the learners to write it down. Compare alternative answer forms (for example 15 minutes past 8, 8.15, a quarter past eight).

Reading scales game

This game assesses the learners' ability to read partially numbered scales to the nearest division or half division.

You will need:

Sets of 36 mass and volume cards made from photocopiable pages 110 and 112 and a set of answer cards.

What to do

- Organise the learners into groups of six. Give each group a set of 36 cards. The dealer should shuffle the cards and deal out six cards to each player.
- The player to the left of the dealer should choose a card from their hand and put it face up on the table. The player with the matching card must place it face up next to the other card. When everyone has checked that the two cards match, the dealer should turn them face down. It is now the turn of the player who played the matching card to play a card of their choice. The game is over when one player has no more cards left in their hand. This player is the winner.

Written assessment

Distribute photocopiable page 114. Ask the learners to read and answer the questions. They should work independently. Questions 7 and 8 require practical measuring equipment. You may want to stagger the order in which different groups do the questions.

Name: _____

Can you solve the problems?

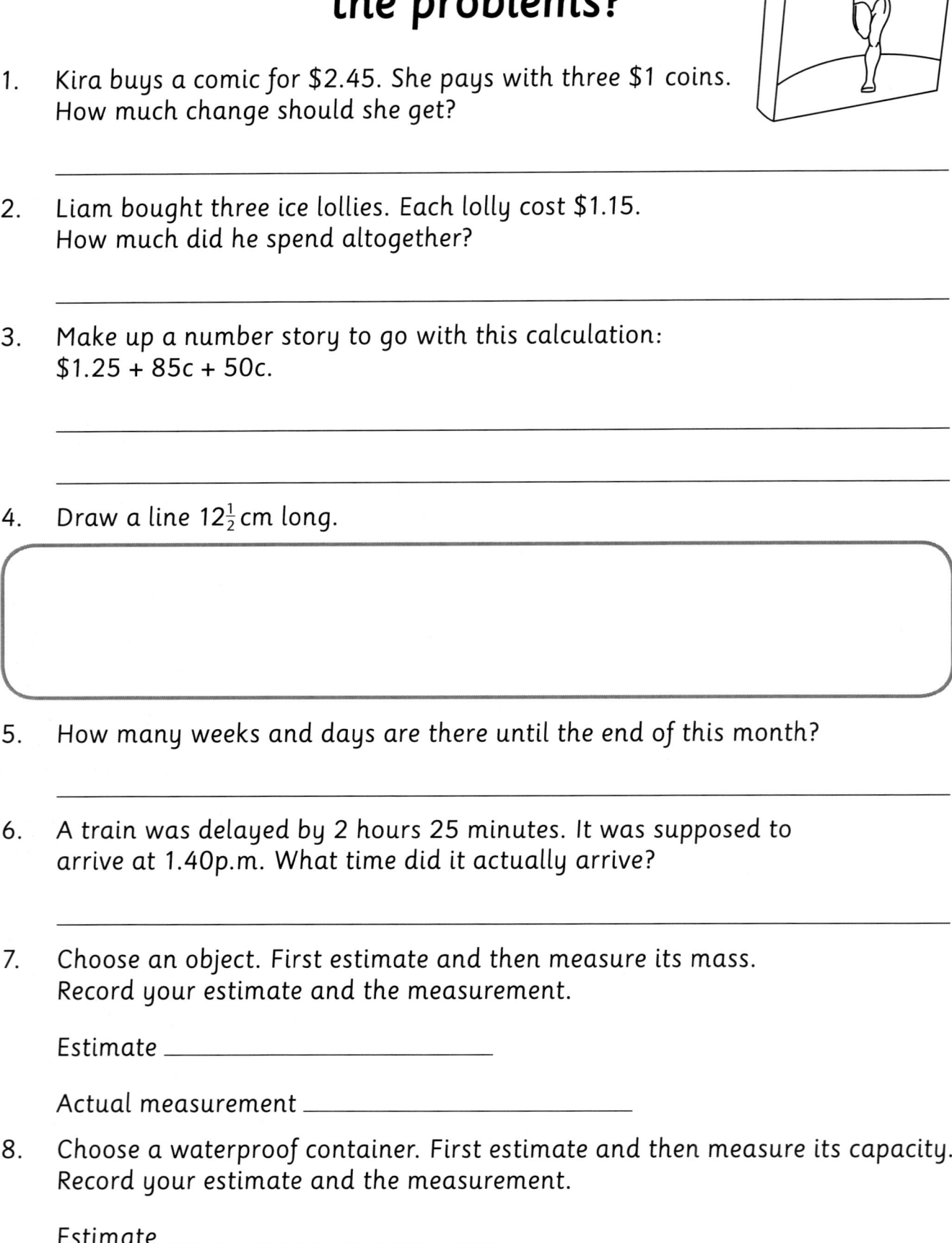

1. Kira buys a comic for $2.45. She pays with three $1 coins.
    How much change should she get?

    _____

2. Liam bought three ice lollies. Each lolly cost $1.15.
    How much did he spend altogether?

    _____

3. Make up a number story to go with this calculation:
    $1.25 + 85c + 50c.

    _____

    _____

4. Draw a line $12\frac{1}{2}$ cm long.

5. How many weeks and days are there until the end of this month?

    _____

6. A train was delayed by 2 hours 25 minutes. It was supposed to
    arrive at 1.40p.m. What time did it actually arrive?

    _____

7. Choose an object. First estimate and then measure its mass.
    Record your estimate and the measurement.

    Estimate _____

    Actual measurement _____

8. Choose a waterproof container. First estimate and then measure its capacity.
    Record your estimate and the measurement.

    Estimate _____

    Actual measurement _____

Cambridge Primary: Ready to Go Lessons for Maths Stage 3 © Hodder & Stoughton Ltd 2013

Unit 2C: Handling data and problem solving

Tally charts and frequency tables

- Use tally charts, frequency tables, pictograms (symbol representing one or two units) and bar charts (intervals labelled in ones or twos). (3Dh2)
- Use ordered lists and tables to help solve problems systematically. (3Ps4)

Photocopiable page 116; counting stick; squared paper; rulers; coloured pencils.

Starter

- Tell the learners that they will be drawing and reading tally charts in today's lesson, and that practising counting in fives will help them do this more easily. Briefly show and discuss the tally chart from photocopiable page 116.
- Practise counting on and back in fives using a counting stick. Start by counting between 0 and 50, and extend to counting past 50. Next, point to a mark on the stick (for example the fourth mark) and say: *Four 5s.* Ask the learners to reply as quickly as they can with '20'. Repeat in a random order.

Main activities

- Pose a question that can be answered by collecting numerical data (whole numbers), for example: *In our class / school, which is the most popular way of travelling to school?* Alternatively, choose a subject that interests the learners, or that links to current events in the school or wider community.
- Ask the learners to predict the answer to the question and explain their reasoning. Ask them what information they will need to collect to answer the question, and how they might collect it.

- Display and discuss the tally chart on photocopiable page 116.
- Organise the learners into groups, asking each group to collect a separate part of the data. Collate the data and create a single tally chart on the board.
- Show the ordered frequency table on photocopiable page 116 and explain how it relates to the data in the tally chart. Ask the learners to create an ordered frequency table from the data in the whole-class tally chart.

Plenary

- Recap the question you asked at the beginning of the lesson. Ask the learners to answer it, using the data they've collected to support their answer.
- Ask the learners to describe anything else they have discovered from examining the data.

Ask the learners:

- What question are we trying to answer? What do you think the answer to the question will be? Why?
- What data are you collecting? How will this data help us to answer the question we're trying to answer?
- What does the data in this chart / table tell you?

Support: To support these learners, organise mixed-ability groupings in the main activity.

Extension: In the final part of the main activity, challenge these learners to represent the data graphically, by drawing a bar chart or a pictogram.

Tally chart and ordered frequency table

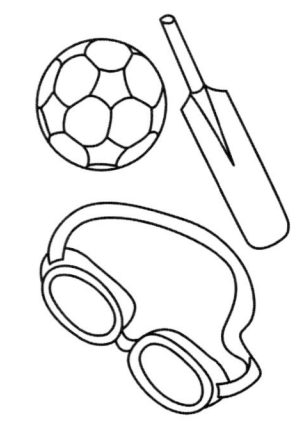

What is the most popular sport in Classes 3, 4 and 5 at Kingston School?

Tally chart

Sport	Tally
Athletics	卌 卌 l
Basketball	卌 卌 ll
Cricket	卌 卌 卌 llll
Cycling	卌 lll
Football	卌 卌 卌 卌 卌 l
Netball	卌 l
Rugby	lll
Swimming	卌 卌 卌 卌 l

Ordered frequency table

Sport	Frequency
Football	26
Swimming	21
Cricket	19
Basketball	12
Athletics	11
Cycling	8
Netball	6
Rugby	3

   Cambridge Primary: Ready to Go Lessons for Maths Stage 3 © Hodder & Stoughton Ltd 2013

Pictograms

- Use tally charts, frequency tables, pictograms (symbol representing one or two units) and bar charts (intervals labelled in ones or twos). (3Dh2)

Counting stick; photocopiable page 118; squared paper; coloured pencils.

Starter

- Practise counting on and back in twos using a counting stick. Start by counting between 0 and 20, and extend to counting past 20. Next, point to a mark on the stick (for example the sixth mark) and say: *Six 2s*. Ask the learners to reply as quickly as they can with '12'.
- Repeat in a random order.

Main activities

- Display a copy of photocopiable page 118. Explain that to make sense, a pictogram must always have a key. Ask questions about the data shown in each pictogram.
- Ask a simple question that can be answered by counting numbers of people (for example: *What is the favourite day of the week in our class?*). Write the days of the week on the board. Next to each day write the number of learners who vote for it as their favourite day.
- Work with the learners to draw a pictogram of the data you have just collected, using a symbol that represents two units. In the example, the categories in the pictogram will be the days of the week, and the data recorded in each category will be the number of learners who voted for that day. The symbol you use will represent two children. You might choose, for example, a stick figure or a smiley face. For any days of the week that get an odd number of votes, the final symbol you draw in that category will need to be a half symbol.

- Ask another question that can be answered by counting numbers of people (for example: *What are our favourite fruits?*). Organise the learners into groups to collect the data they need in order to answer the question. Ask them to record the data they collect in a pictogram, remembering to include a key.

Plenary

- Recap the question you asked the learners to find out the answer to. Ask the learners to answer it, using the data they've collected to support their answer.
- Ask the learners to describe anything else they have discovered from examining the data.

Ask the learners:

- What does each symbol represent in this pictogram? How do you know?
- Tell me something that this pictogram shows.

Support: Group these learners together and work with them to collect, record and interpret the data.

Extension: Ask these learners to devise their own question that can be answered by collecting numerical data, and then answer it.

Pictograms

Birds in the playground

Key: 🐦 = 1 bird

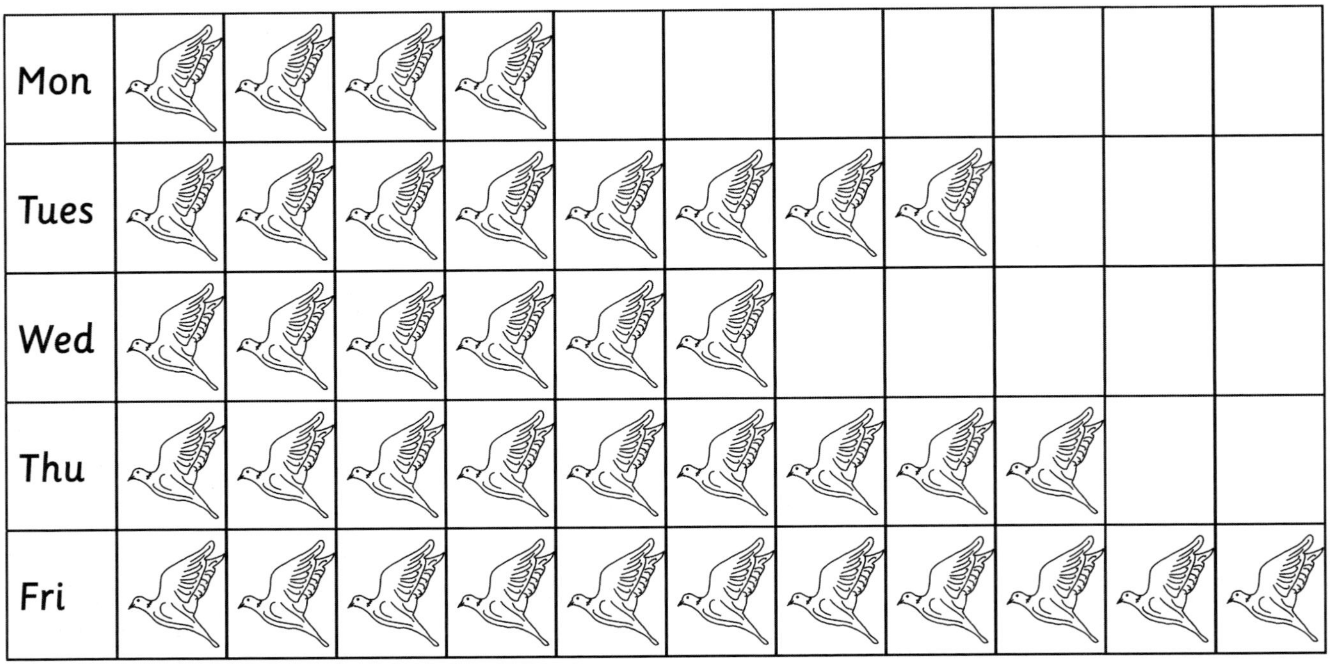

Mon	🐦	🐦	🐦	🐦							
Tues	🐦	🐦	🐦	🐦	🐦	🐦	🐦	🐦			
Wed	🐦	🐦	🐦	🐦	🐦	🐦					
Thu	🐦	🐦	🐦	🐦	🐦	🐦	🐦	🐦	🐦		
Fri	🐦	🐦	🐦	🐦	🐦	🐦	🐦	🐦	🐦	🐦	🐦

Goals in a football season

Key: ⚽ = 2 goals scored

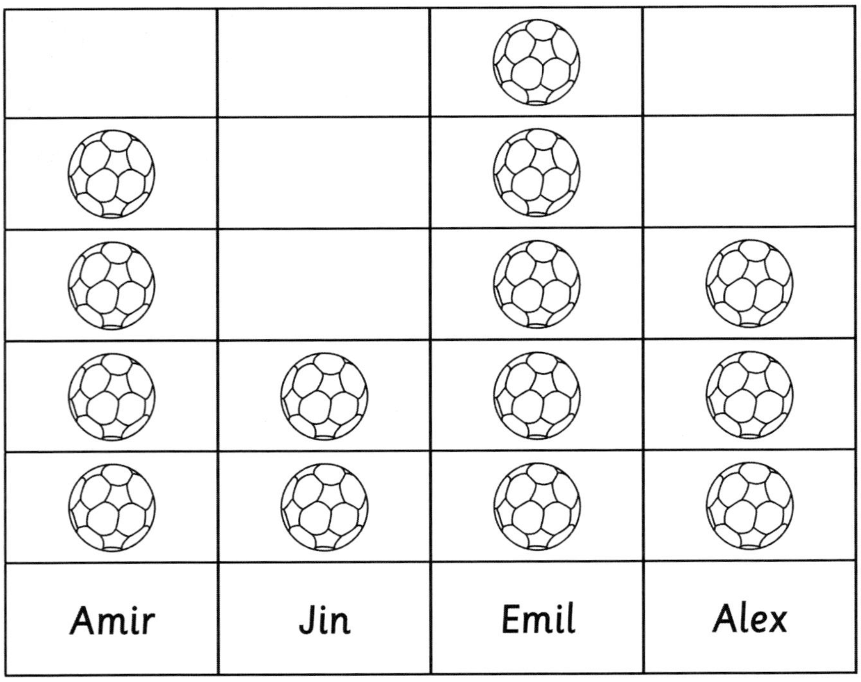

		⚽	
⚽		⚽	
⚽		⚽	⚽
⚽	⚽	⚽	⚽
⚽	⚽	⚽	⚽
Amir	Jin	Emil	Alex

Bar charts

● Use tally charts, frequency tables, pictograms (symbol representing one or two units) and bar charts (intervals labelled in ones or twos). (3Dh2)

Photocopiable page 120; squared paper; rulers; coloured pencils.

Starter

• On the board sketch a blank number line with 20 divisions. Practise counting on from zero and back again, first in twos (to 40) and then in fives (to 100).

• Draw a number of dots on the board (between 5 and 20). Say: *Each dot represents 2. What number do all the dots together represent?* Repeat the question with each dot representing 5. Repeat for different numbers of dots.

Main activities

• Display a copy of photocopiable page 120. Ask questions about the data shown in each bar chart.

• Ask a simple question that can be answered by collecting discrete numerical data (for example: *Which is the most common shoe size in this class?*). Ask the learners to predict what they think the answer to the question will be, and explain their reasoning. Collect the data together and record the data in a tally chart.

• Work with the learners to draw a bar chart of the data, using a scale marked in intervals of ones or twos.

• Ask another question that can be answered by collecting discrete numerical data (for example: *Which is the most popular school subject in this class / school?*)

• Organise the learners into groups to collect the data they need in order to answer the question. Ask the learners first to record the data they collect in a tally chart, and then in a bar chart.

Plenary

• Recap the question you asked. Ask the learners to answer it, using the data they've collected to support their answer.

• Ask the more-able learners who have done the extension activity to compare and contrast the four different chart types (tally chart, frequency table, pictogram and bar chart).

Ask the learners:

● Tell me something that this bar chart shows.

● Tell me something that this tally chart shows.

● Draw a tally chart to show the number of learners in your class who came to school each day last week.

● Draw a bar chart to show the same data.

Support: Group these learners together and work with them to collect, record and interpret the data.

Extension: Ask these learners to draw a pictogram and a frequency table of the data for which they've already drawn a tally chart and a bar chart. Ask them to discuss the similarities and differences among these charts.

Bar charts

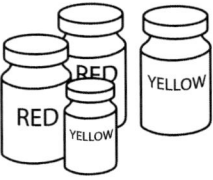

Favourite colours in Classes 4 and 5

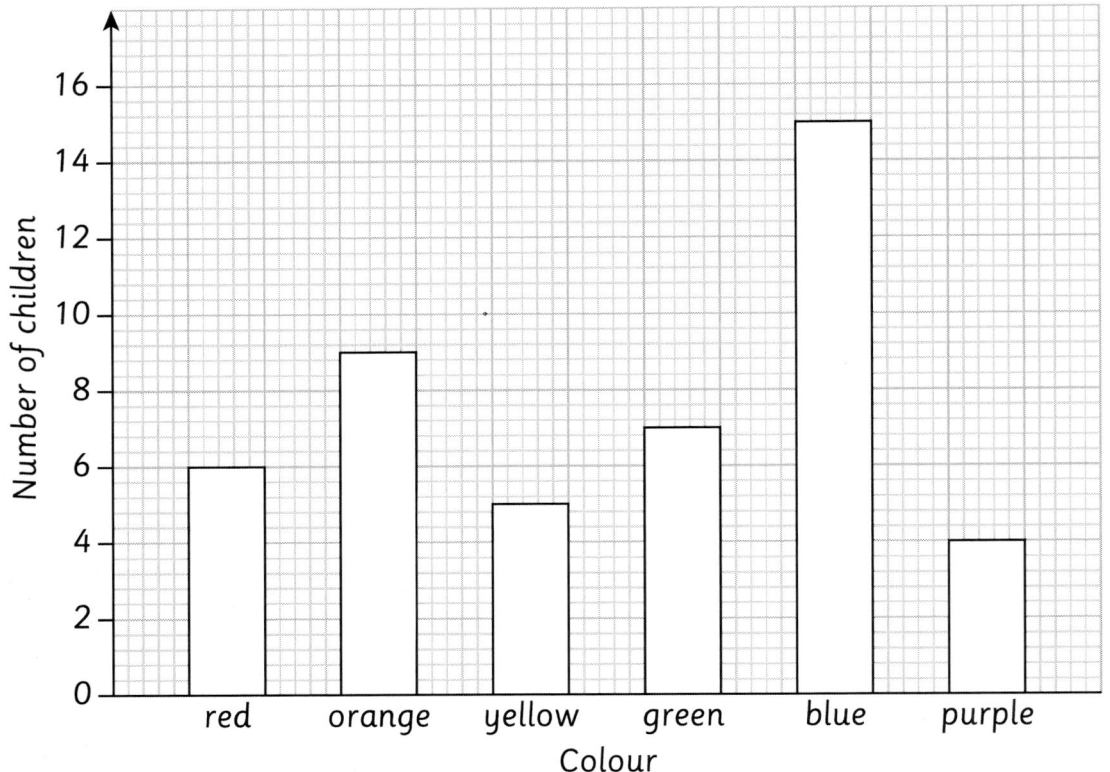

Sales at Ned's ice cream bar

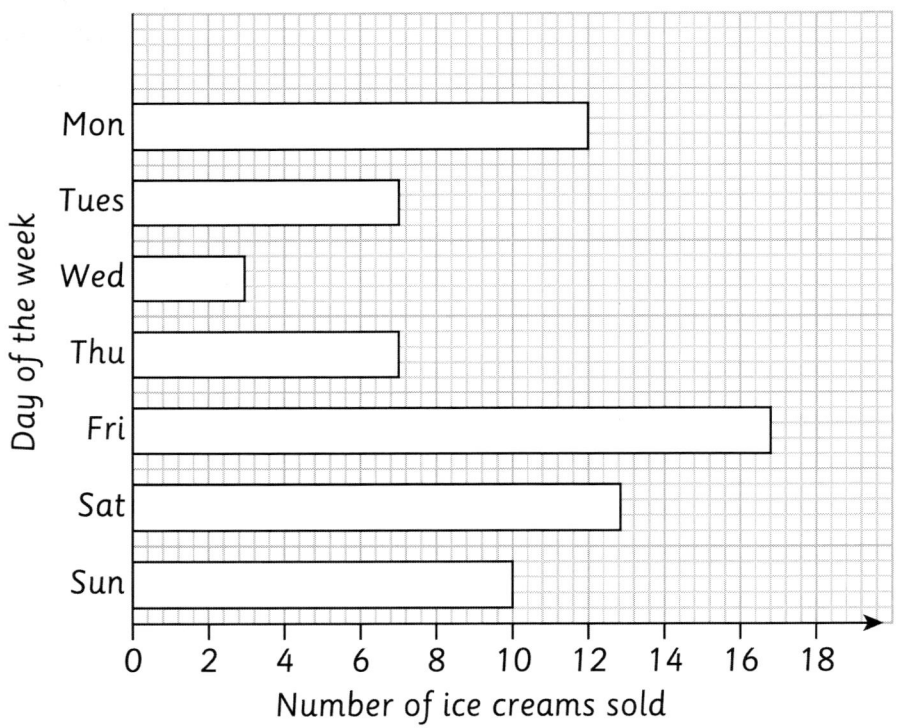

Cambridge Primary: Ready to Go Lessons for Maths Stage 3 © Hodder & Stoughton Ltd 2013

Carroll diagrams

Learning objectives

● Use Venn or Carroll diagrams to sort data and objects using two criteria. (3Dh3)

Resources

Pencils; plain paper; selection of everyday objects; photocopiable page 122; 2D shape templates.

Starter

● Organise the learners into pairs and give each pair paper and pencils.

● Sort a selection of everyday objects into two groups according to a secret criterion (for example made of wood / not made of wood). Label the groups A and B. Ask the learners to write a description of each group. Repeat the activity, asking a volunteer to sort the objects into two groups.

● On the board write a selection of random numbers. Ask the learners to suggest a way of sorting the numbers into two groups.

Main activities

● Display a copy of photocopiable page 122 and use it to explain how a Carroll diagram works.

● Draw an unlabelled Carroll diagram on the board.

● Call up about half a dozen volunteers to the front of the class. Ask the learners to suggest various ways of sorting the volunteers into two groups, and record all the valid suggestions (for example wearing a jumper / not wearing a jumper, straight hair / not straight hair). Label the columns in the diagram with one way of sorting the volunteers, and label the rows with another. Ask the learners whereabouts in the diagram each volunteer's name should be written. Write in the names.

● Organise the learners into groups of about six. Ask them to choose two ways of sorting the people in their group into two groups, and to record their sorting in a Carroll diagram.

Plenary

● On the board draw a four-box Carroll diagram. Label the columns 'Birthday January to June / Birthday July to December' and the rows 'Walk to school / Don't walk to school'. Ask the learners to put their hands up when you point to 'their' box.

Success criteria

Ask the learners:

● Where does [a learner's name] belong in this Carroll diagram? Why?

● Draw a Carroll diagram to sort these objects, sorting them into two groups in two different ways.

Ideas for differentiation

Support: For the final main activity, group these learners together and work with them.

Extension: Ask these learners to sort 2D shape templates according to their own criteria, and record the information in a Carroll diagram.

A Carroll diagram

	At least 1 right angle	No right angles
4 sides		
Not 4 sides		

Cambridge Primary: Ready to Go Lessons for Maths Stage 3 © Hodder & Stoughton Ltd 2013

Venn diagrams

Learning objectives

● Use Venn or Carroll diagrams to sort data and objects using two criteria. (3Dh3)

Resources

Times table charts of the 3 and 4 times tables; sticky notes; photocopiable page 124; 2D shapes (such as square, rectangle, trapezium, equilateral triangle, isosceles triangle, rhombus, circle, pentagon, hexagon, octagon); A3 paper; pencils.

Starter

- Display the chart of the 3 times table and practise chanting it. Each time you chant, cover up a couple of products with sticky notes until all the products are covered up.
- Repeat for the 4 times table.

Main activities

- Display a copy of photocopiable page 124 and use it to explain how a Venn diagram works. Explain that this diagram shows one way of sorting the whole numbers from 1 to 30.
- Draw a two-circle Venn diagram on the board. Label one circle 'symmetrical' and the other 'more than four sides'. Display a selection of 2D shapes, at least one of which belongs in each region of the diagram. Ask the learners to place each shape in the correct region of the diagram.
- Organise the learners into groups. Give each group a selection of 2D shapes and some A3 paper. Ask each group to draw a large Venn diagram to sort the shapes, and put each shape in the correct region of the diagram. The learners should make their own choices about the criteria they use to sort the shapes, and keep them secret from other groups.
- Ask the learners to cover up the labels on their Venn diagram (for example using sticky notes), and then ask the learners to view each other's diagrams, and try to work out the label for each circle.

Plenary

- On the board draw a two-circle Venn diagram. Label the circles 'Likes pizza' and 'Has at least one pet'. Ask the learners to put their hands up when you point to 'their' region of the diagram.

Success criteria

Ask the learners:

● Where does [a learner's name] belong in this Venn diagram? Why?
● Draw a Venn diagram to sort these objects.
● Can you label the circles on this Venn diagram?
● What does this Venn diagram show?

Ideas for differentiation

Support: In the final main activity, give these learners a pre-drawn and pre-labelled two-circle Venn diagram.

Extension: Ask these learners to draw a second Venn diagram in which they sort their shapes in a different way.

A Venn diagram

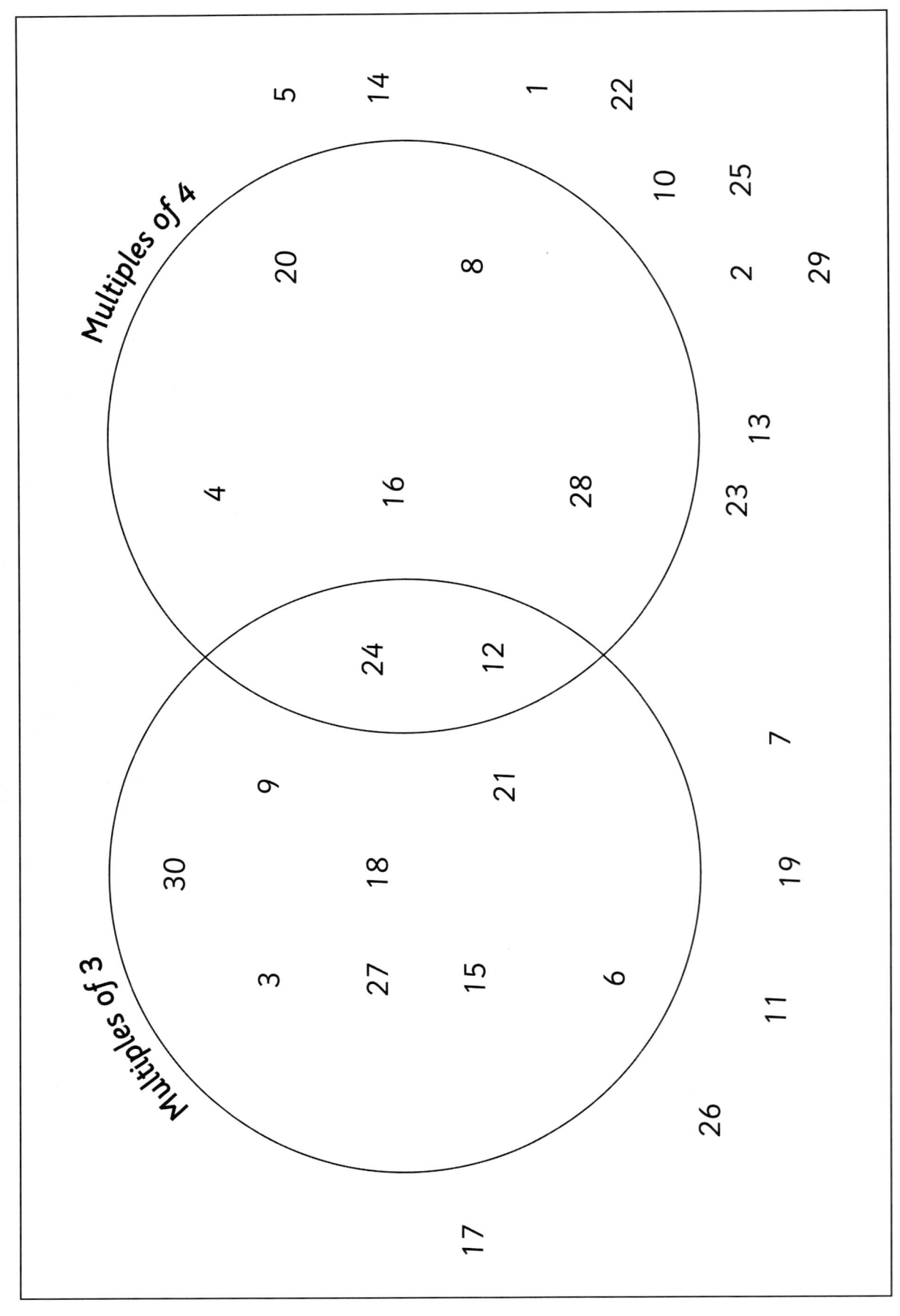

Multiples of 4

Multiples of 3

5 14 1 22

10 25

20 8

2 29

4 16 28 13

23

24 12

9 21 7

30 18 19

3 27 15 6 11

26

17

Cambridge Primary: Ready to Go Lessons for Maths Stage 3 © Hodder & Stoughton Ltd 2013

Data handling project

Learning objectives

- Answer a real-life question by collecting, organising and interpreting data, e.g. investigating the population of mini-beasts in different environments. (3Dh1)
- Use tally charts, frequency tables, pictograms (symbol representing one or two units) and bar charts (intervals labelled in ones or twos). (3Dh2)

Resources

Counting stick; 0 to 9 number fans (see photocopiable page 24); photocopiable pages 116, 118, 120 and 126; clipboards; copies of the class register; tape measures; stopwatches; various brands of cola; cups.

Starter

- Using a counting stick count on from zero in twos, fives and tens and back again.
- Give each learner a 0 to 9 number fan. Call out a fact from the 2, 5 or 10 times table. Ask the learners to make the answer on the number fan as quickly as they can and show it to you. Avoid calling out *ten times ten*, as it is not possible to make 100 on a number fan.

Main activities

- Display a copy of photocopiable page 126, which poses a variety of questions that can be answered by collecting numerical data. For each question, ask the learners to predict the answer and explain their reasoning. Ask: *What data would you need to collect in order to answer the question? How might you collect it?*
- Revise tally charts, frequency tables, pictograms and bar charts as ways of organising and presenting data. You could do this by displaying copies of photocopiable pages 116, 118 and 120.
- Organise the learners into groups. Ask each group to choose one of the questions on photocopiable page 126 to answer. Each group should decide for themselves how they will collect the data and how they will present it.

Plenary

- Ask a representative from each group to say which question they chose to answer, explain how they collected the data and display the graph they drew.
- Ask each group: *Have you found out enough information in order to answer the question you asked? If so, what's the answer?*

Success criteria

Ask the learners:

- Which question did you choose?
- What data did you collect and how did you collect it?
- How did you organise and present the data?
- What does the data you collected show?

Ideas for differentiation

Support: Organise the learners into mixed-ability groups, so that these learners are supported by other group members.

Extension: Ask these learners to ask and answer their own follow-up question to the initial question they investigated.

Questions to answer by collecting and organising data

1. What is the most popular colour of car?

2. Do children have better eyesight than adults?

3. What is the most popular sandwich filling in our class?

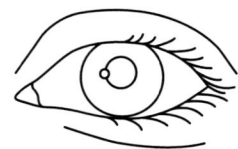

4. Can people identify a brand of cola from taste alone?

5. Are there more absences on some days of the week than on others?

June

M	T	W	Th	F	Sa	Su
	1	2	3	4	5	6
7	8	9	10	11	12	13
14	15	16	17	18	19	20
21	22	23	24	25	26	27
28	29	30				

6. Are people with longer legs faster runners?

Cambridge Primary: Ready to Go Lessons for Maths Stage 3 © Hodder & Stoughton Ltd 2013

Unit assessment

Display a copy of photocopiable page 128, and ask the following questions:

● What type of chart is this?

● What might the title of this chart be?

● What might be the labels here (in the left-hand square of each row)?

● What might be the missing numbers in the key?

● What questions could you ask about this data?

Summative assessment activities

Observe the learners while they take part in these activities. You will quickly be able to identify those who appear to be confident and those who may need additional support.

Drawing a bar chart

This activity assesses the learners' ability to interpret tally charts and draw bar charts.

You will need:

Squared paper; rulers; pencils; erasers; coloured pencils.

What to do

● Draw a tally chart on the board. You could use or adapt the chart below.

Favourite foods in Classes 3 and 4	
Favourite food	**Number of children**
Pizza	Жℋℋ Жℋℋ II
Enchillada	Жℋℋ I
Tortilla	Жℋℋ Жℋℋ Жℋℋ IIII
Chocolate	Жℋℋ III
Watermelon	Жℋℋ Жℋℋ IIII
Mango	Жℋℋ II
Fries	III
Tacos	Жℋℋ Жℋℋ

● Ask the learners to draw a bar chart of the data shown in the tally chart.

Sorting data

This activity assesses the learners' ability to organise data into Carroll diagrams.

You will need:

Ten-sided dice (0 to 9).

What to do

● Draw a Carroll diagram on the board with columns labelled 'odd' and 'even' and rows labelled 'digits total less than 10' and 'digits total 10 or more'.

● Give each learner a ten-sided dice. Ask the learners to use the dice to generate two-digit numbers, and then write each number in the correct region of the diagram.

Distribute photocopiable page 128, and ask the learners to complete it. Remind them that they need to fill in the type of chart, give the chart a title, label the boxes on the vertical axis, and complete the key. Ask each learner to write five questions about the data shown on the chart, and give the questions, together with their completed chart, to a friend. The friend should answer the questions in writing.

Ask those who have finished to draw their own pictogram, showing any of the sets of data they have collected during the unit, but which they have not yet represented as a pictogram.

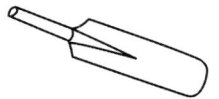

Name: _____

Mystery chart

1. What type of chart is shown below? _____

2. What title could you give the chart? _____

3. Now try to complete the key.

4. How could the rows be labelled? Add your labels.

5. What questions could you ask about the data?

_____

_____

_____

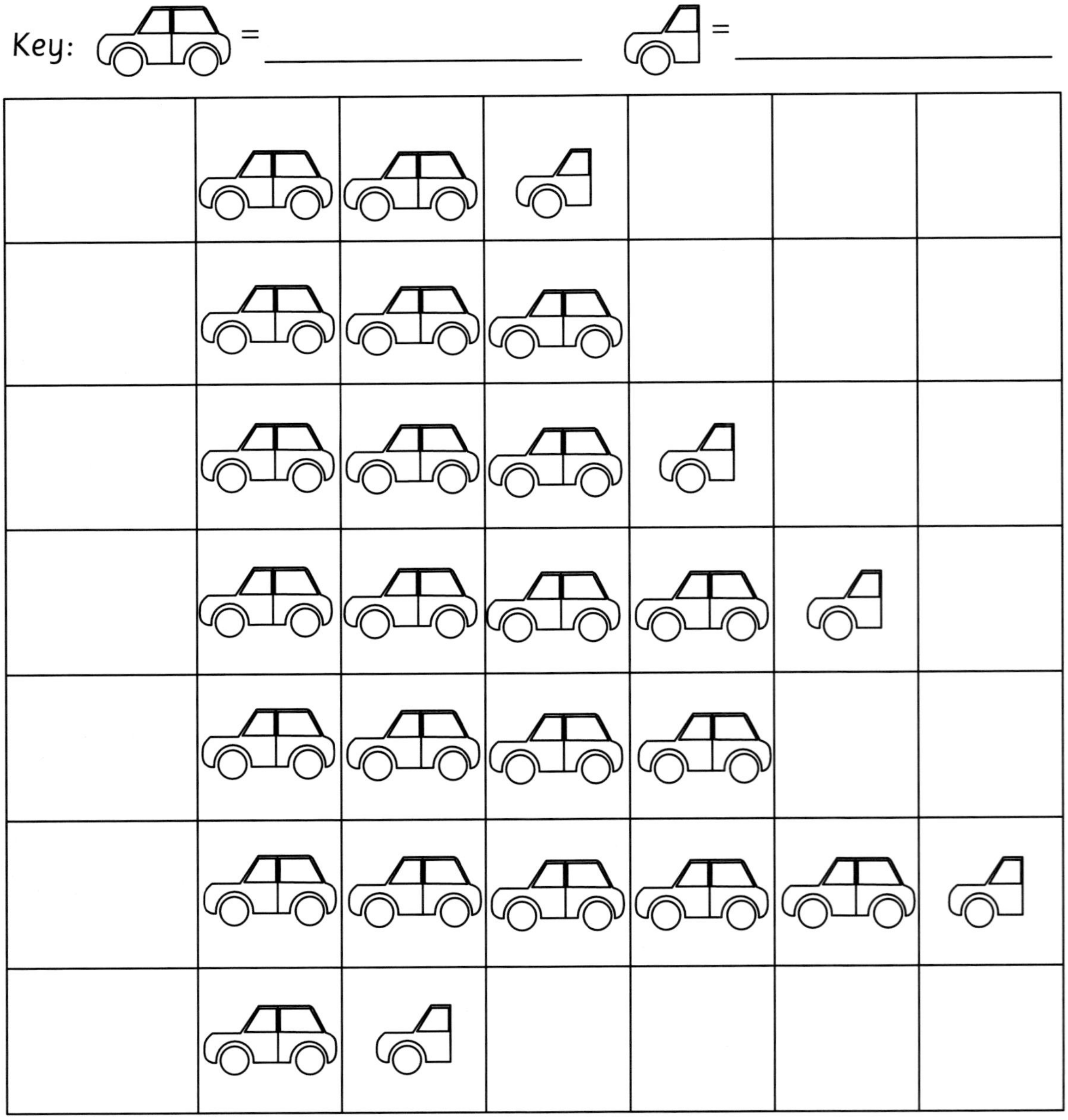

Cambridge Primary: Ready to Go Lessons for Maths Stage 3 © Hodder & Stoughton Ltd 2013

Unit 3A: Number and problem solving

Place value 5

Starter

- On the board write a number sequence that counts on or back in steps of 2, 3, 4, 5, 10 or 100, for example 32, 37, 42, 47, 52; 107, 105, 103, 101, 99; 238, 338, 438, 538, 638; 72, 75, 78, 81, 84; 125, 115, 105, 95, 85; 9, 13, 17, 21, 25.

- Ask the learners to write the next three numbers in the sequence, and then share their answers. Ask them to describe the number sequence (for example 'You add five each time', 'It goes up in threes', 'Each number is four less than the one before'). Repeat for several number sequences.

Main activities

- Organise the learners into pairs and hand out paper and pencils. Draw this place value chart on the board:

Thousands	Hundreds	Tens	Units

- Write a three-digit number in the chart. Ask the learners to write the number that is 10 / 20 / 30 ... more or less. Ask: *Which digit did you need to change? How did you work out how many you needed to change the digit by?* Include some examples that involve crossing the hundreds boundary (for example: *What is 40 less than 231?*)

- Ask the learners to investigate the statement: 'When you add a multiple of 100 to a three-digit number, the only digit that changes is the first one'. The learners may use calculators for this activity.

- Discuss examples of calculations that satisfy the statement (for example calculations involving whole numbers only that total less than 1000) and examples of calculations that do not satisfy it (for example calculations that total more than 999 and calculations that involve three-digit decimals). Work together to rewrite the statement so that it holds true for all numbers.

Plenary

- Enlarge several copies of photocopiable page 130 onto A3 paper (or onto A3 card and laminate). Display the function machine and write numbers in the input and output boxes, the difference between them being a multiple of 10 or 100. Ask the learners to tell you the missing function (for example '+ 60' or '– 300'). Repeat several times.

Function machine

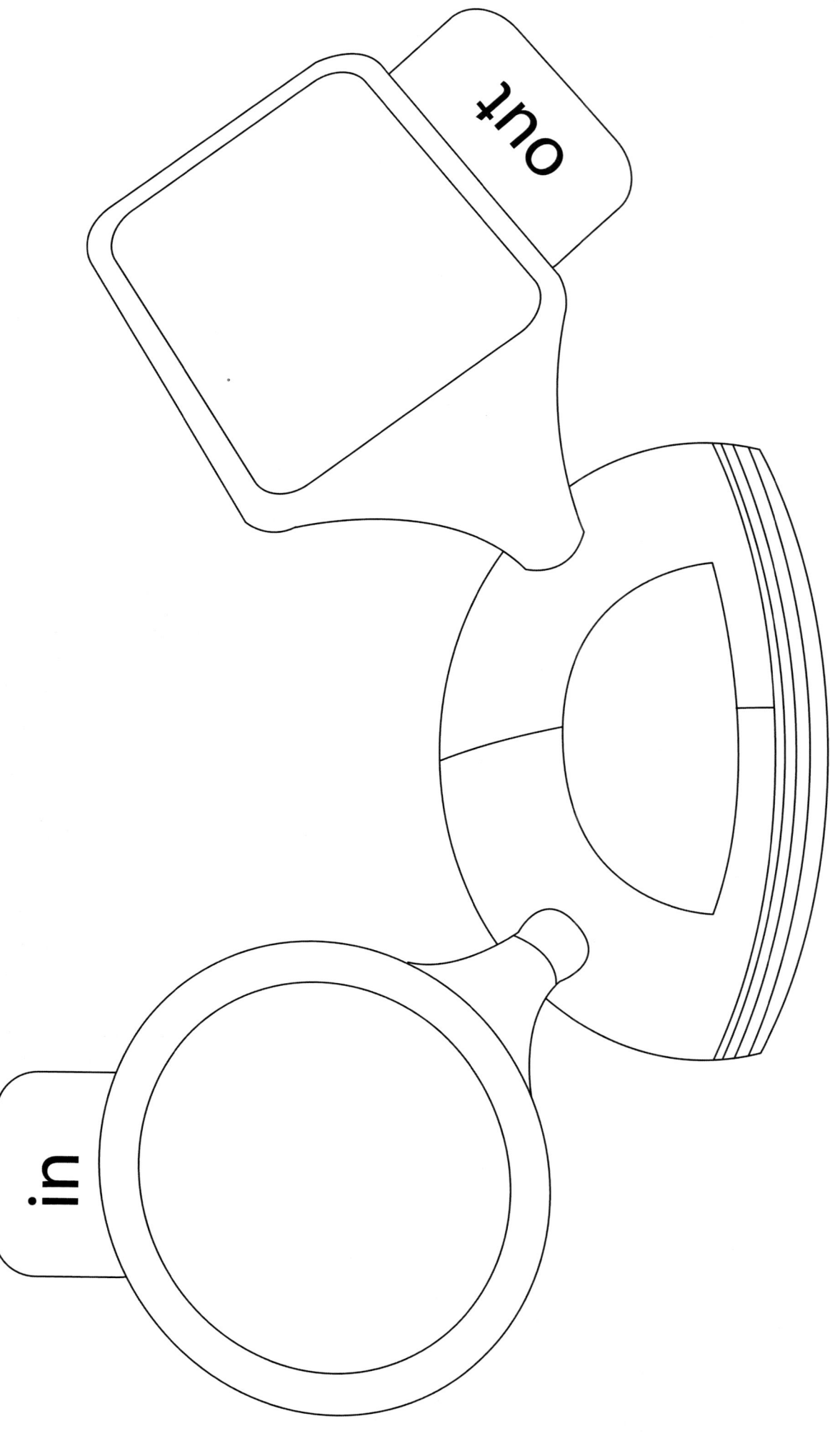

in

out

Comparing and ordering numbers 2

Learning objectives

- Multiply two-digit numbers by 10 and understand the effect. (3Nn7)
- Compare three-digit numbers, use < and > signs, and find a number in between. (3Nn11)
- Order two- and three-digit numbers. (3Nn12)

Resources

Two large sets and plenty of small sets of 0 to 9 digit cards; number lines made from photocopiable page 132; pencils; plain paper.

Starter

- Shuffle together the two large sets of 0 to 9 digit cards. Draw two cards and hold them up to make a two-digit number (for example 31). Ask the learners to multiply the number by 10 and write the answer (for example 310). Repeat for several two-digit numbers. Next, draw two cards in secret and make a secret two-digit number with them (for example 47). Write on the board the number you made multiplied by 10 (for example 470). Ask the learners to work out which number you made with the cards. Repeat several times.

Main activities

- Revise the greater than and less than signs.
- Shuffle the digit cards. Draw the top six cards to make two three-digit numbers. Ask the learners to insert the correct symbol between the numbers (< or >). Ask: *Which is the most important digit when comparing the numbers? Why?* Give some examples in which the hundreds digit is the same in both numbers and ask the same question. Repeat for a pair of numbers in which both the hundreds and tens digits are the same. Finally, ask the learners to give a number between the two numbers.
- Organise the learners into pairs and give each pair two small sets of 0 to 9 digit cards.
- Write ten two- and three-digit numbers on the board, asking pairs to order them from largest to smallest. Ask the learners to write their own list of ten numbers, and give it to their partner to order.

Plenary

- Using the two large sets of 0 to 9 digit cards, draw the top eight cards to make two four-digit numbers. Ask the learners to insert the correct symbol between the numbers.
- Ask: *Which is the most important digit when comparing these two numbers? Why? Will the most important digit be the same for all pairs of four-digit numbers?*

Success criteria

Ask the learners:

- Is this number sentence correct? 834 > 843 Why? / Why not?
- Write a number sentence using the symbol <.
- Write a number between 598 and 602.
- Write these numbers in order from largest to smallest: 99, 154, 408, 150, 48, 614, 432, 91, 56, 641.

Ideas for differentiation

Support: To help these learners with comparing and ordering numbers, give them 0 to 1000 number lines made from enlarging photocopiable page 132 onto A3 card.

Extension: In the final main activity, challenge these learners to write lists of four-digit numbers for their partners to order.

0 to 1000 number line

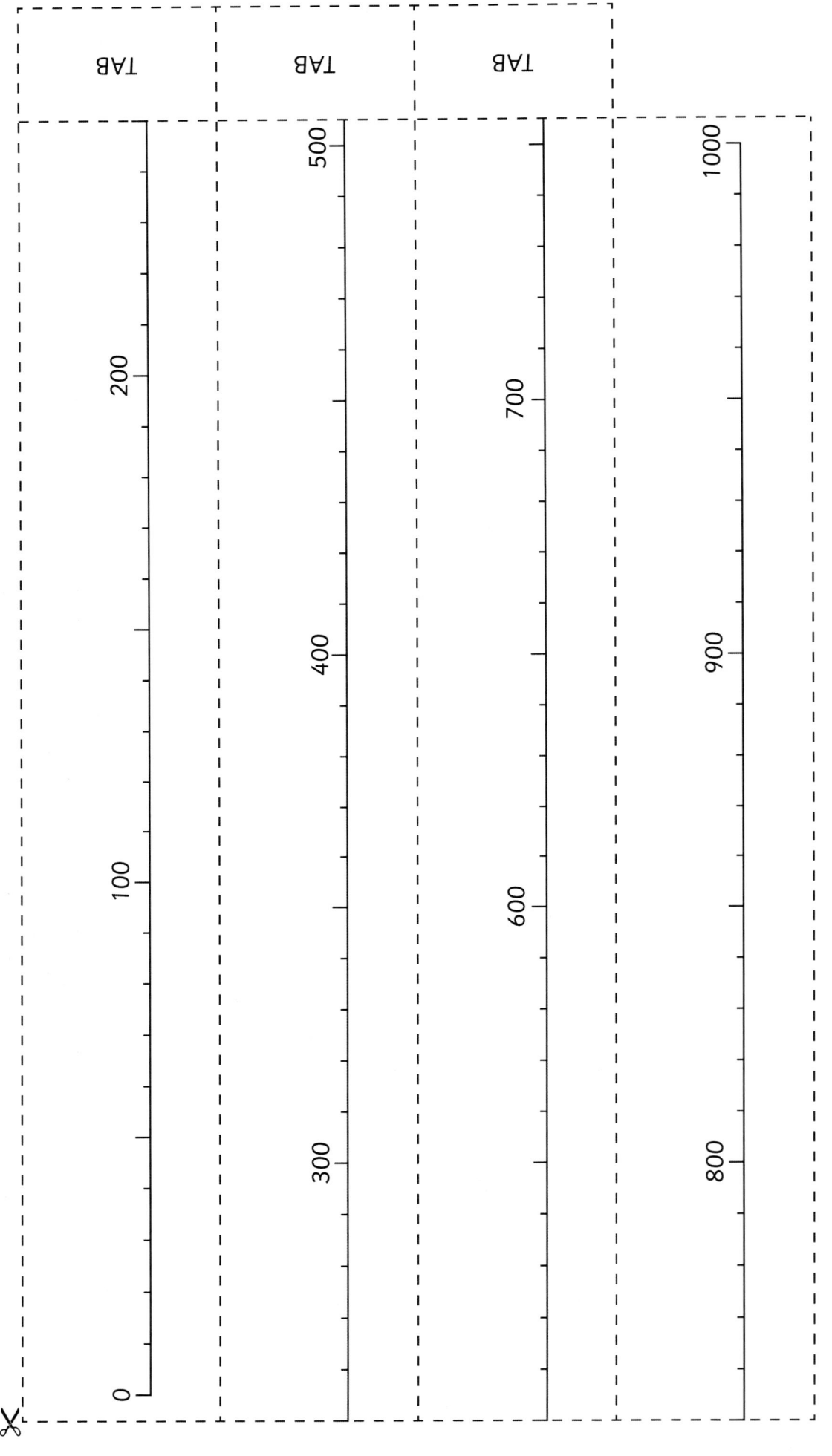

Estimating and rounding 3

Learning objectives

- Round two-digit numbers to the nearest 10 and round three-digit numbers to the nearest 100. (3Nn8)
- Give a sensible estimate of a number as a range (e.g. 30 to 50) by grouping in tens. (3Nn13)
- Estimate and approximate when calculating, and check working. (3Pt10)
- Make a sensible estimate for the answer to a calculation, e.g. using rounding. (3Pt11)

Resources

A group of objects for estimating, e.g. a box of paper clips, a box of matches, a spoonful of rice, a pile of paper; laminated (if possible) place value charts made from photocopiable page 74; dry wipe pens; cloths; photocopiable page 134; number lines; multiplication grids.

Starter

- Display the group of objects for estimating. Ask the learners to work with a partner to estimate the number of objects.
- Collect estimates and discuss methods used. Reveal the actual answer.

Main activities

- Organise the learners into pairs and give each pair a laminated copy of a place value chart made from photocopiable page 74, a dry wipe pen and a cloth.
- On the board draw a place value chart like the one on photocopiable page 74. Write a two-digit number in the place value chart, asking the learners to round it to the nearest 10 and write the answer on their place value chart. Repeat for several two-digit numbers, including some that end in a 5. Discuss the digit rules for rounding to the nearest 10.
- Repeat for rounding three-digit numbers to the nearest 100.

- Display a copy of photocopiable page 134. Demonstrate how to use rounding to the nearest 10 (for two-digit numbers) and nearest 100 (for three-digit numbers) to estimate the answer to a calculation.
- Ask pairs of learners to make and record estimates for the remainder of the calculations. Discuss the estimates.
- Ask the learners to perform the calculations on photocopiable page 134, checking their answers by comparing them to their estimates.

Plenary

- Discuss the answers to the questions on photocopiable page 134.
- Compare estimates to answers. Ask the learners to identify features of calculations in which the estimate is very close to the actual answer and those in which it is not so close.

Success criteria

Ask the learners:

- Estimate the number of pages in this book / sweets in this packet. How did you reach your estimate?
- Can you name a number that becomes 900 when you round it to the nearest 100?
- Estimate the answer to 9×54. How did you work it out?

Ideas for differentiation

Support: The calculations on photocopiable page 134 are arranged in order of difficulty. Ask these learners to attempt only the first five questions, and give them practical calculation tools (for example number lines, multiplication grids, and so on).

Extension: Ask these learners to devise a word problem to go with each calculation.

Using rounding to make estimates

1.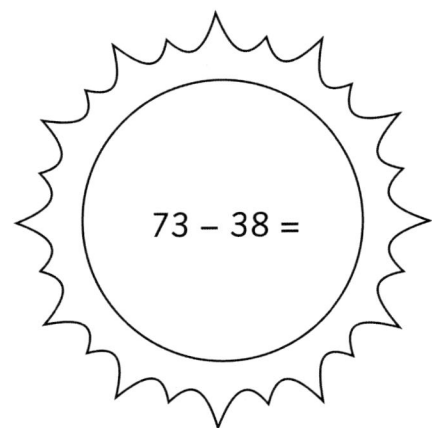

$73 - 38 =$

2.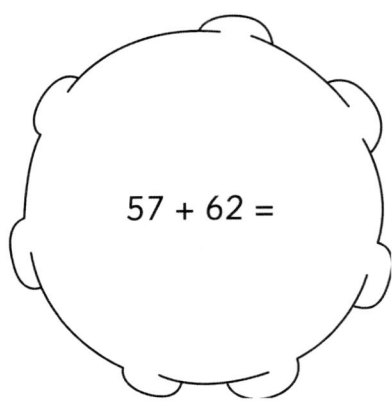

$57 + 62 =$

3.

$29 \times 3 =$

4.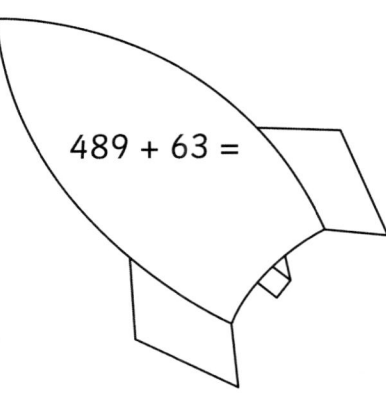

$489 + 63 =$

5.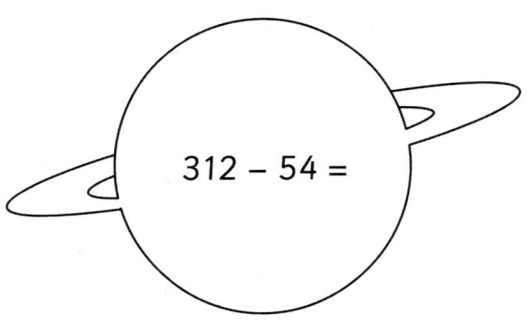

$312 - 54 =$

6.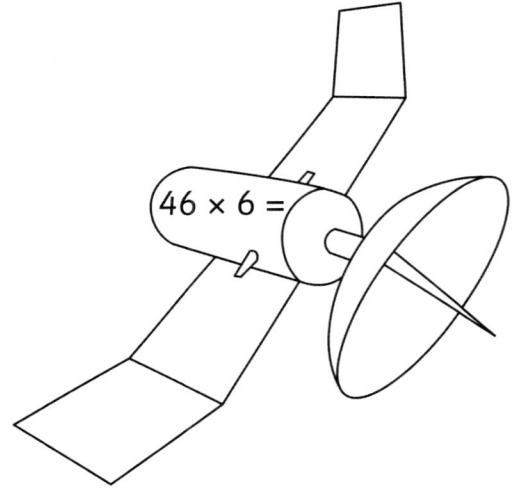

$46 \times 6 =$

7.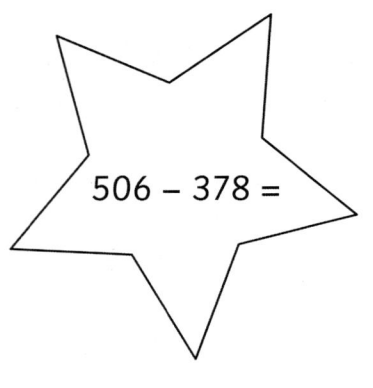

$506 - 378 =$

8.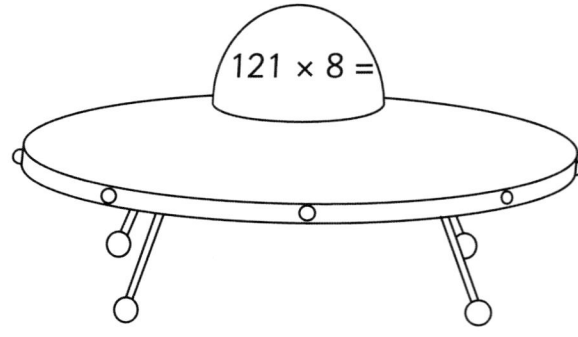

$121 \times 8 =$

Addition and subtraction facts 4

Learning objectives

● Know the following addition and subtraction facts: multiples of 100 with a total of 1000; multiples of 5 with a total of 100. (3Nc2)

Resources

Counting stick; pencils; plain paper; cards made from photocopiable page 136; copies of a crib sheet showing pairs of multiples of 5 that total 100 and pairs of multiples of 100 that total 1000.

Starter

- Using a counting stick, count on from zero in hundreds to 1000 and back again.

- Point to one of the divisions on the counting stick. Ask: *What number is this? How many more to make 1000?* Repeat for other divisions in a random order.

- Put the counting stick away. Say a multiple of 100 (for example two hundred). Ask the learners to say how many more to make 1000 (for example eight hundred). Keep the pace brisk.

Main activities

- Before the lesson make a crib sheet listing pairs of multiples of 5 that total 100 (5 + 95 = 100, 10 + 90 = 100, 95 + 5 = 100, and so on) and pairs of multiples of 100 that total 1000 (100 + 900 = 1000, 200 + 800 = 1000, 900 + 100 = 1000, and so on). Make enough copies for one per group in the second main activity, and one for each learner in groups requiring support.

- Hand out paper and pencils. Hold up one card at a time from a set of cards made from photocopiable page 136. Ask the learners to write the missing number. Keep the pace as brisk as possible.

- Ask the learners to play the following game in groups of four to six: Give each group a set of cards made by enlarging photocopiable page 136 onto A3 card and cutting out. One learner should shuffle the cards and put them in a pile face down. This learner must not play, but act as judge and have a copy of the pairs to 100 and pairs to 1000 crib sheet. The other learners should take it in turns to turn over the top card. The first learner to say the correct missing number wins the card. The winner is the learner with the most cards when there are no more cards left in the deck. Then play again, with a different learner shuffling the cards.

Plenary

- Ask two volunteers to come to the front of the class. Say a multiple of 5 between 5 and 95. The players must say the complement to 100 as quickly as they can. The player who answers first carries on playing. Replace the other player with a new volunteer.

Success criteria

Ask the learners:

● Complete these number sentences:

$\square$ + 35 = 100
70 + $\square$ = 100
100 − $\square$ = 45
$\square$ + 200 = 1000
1000 − $\square$ = 700
1000 − 400 = $\square$

Ideas for differentiation

Support: Group these learners together for the game. Provide each learner with a copy of the crib sheet. Act as dealer in the game.

Extension: Ask these learners to use knowledge of pairs of multiples of 5 that make 100 to derive pairs of multiples of 5 that make 200.

Addition and subtraction cards

$100 - \boxed{} = 95$	$100 - \boxed{} = 90$	$100 - 75 = \boxed{}$	$100 - 70 = \boxed{}$
$100 - \boxed{} = 55$	$100 - \boxed{} = 50$	$100 - 35 = \boxed{}$	$100 - 30 = \boxed{}$
$100 - \boxed{} = 15$	$100 - \boxed{} = 10$	$5 + \boxed{} = 100$	$\boxed{} + 85 = 100$
$\boxed{} + 80 = 100$	$\boxed{} + 65 = 100$	$\boxed{} + 60 = 100$	$45 + \boxed{} = 100$
$40 + \boxed{} = 100$	$\boxed{} + 25 = 100$	$\boxed{} + 20 = 100$	$1000 - 100 = \boxed{}$
$1000 - 500 = \boxed{}$	$1000 - \boxed{} = 600$	$1000 - 900 = \boxed{}$	$1000 - \boxed{} = 200$
$300 + \boxed{} = 1000$	$\boxed{} + 400 = 1000$	$700 + \boxed{} = 1000$	$\boxed{} + 800 = 1000$

Addition and subtraction strategies 3

Learning objectives

- Add and subtract pairs of two-digit numbers. (3Nc14)
- Choose appropriate mental strategies to carry out calculations. (3Pt1)
- Check the results of adding two numbers using subtraction, and several numbers by adding in a different order. (3Pt4)
- Check subtraction by adding the answer to the smaller number in the original calculation. (3Pt5)
- Explain a choice of calculation strategy and how the answer was worked out. (3Ps2)

Resources

Photocopiable page 138; number lines; place value blocks.

Starter

- On the board write $67 + \square = 100$. Ask the learners to find the missing number. Discuss how to use knowledge of multiples of 5 that make 100 to work out the answer, for example knowing $65 + 35 = 100$ lets you work out, by compensation, that $67 + 33 = 100$. Repeat for several similar calculations totalling 100.

Main activities

- On the board write the first question from photocopiable page 138, asking the learners to find the missing number. Discuss calculation strategies used (for example counting on / back on a number line; partitioning and recombining; using compensation from adding / subtracting a multiple of 10, and so on). Explain that different calculations may suggest different methods. Ask the learners to check their answers by using the inverse operation (for example checking addition using subtraction and checking subtraction using addition). Repeat for a couple more questions from photocopiable page 138.

- Organise the learners into pairs and give each pair photocopiable page 138. Ask the learners to find the missing numbers, recording their workings.

Plenary

- Ask the learners to explain their choices of calculation strategy and describe how they worked out the answers.

Success criteria

Ask the learners:

- Can you write an equation that has a missing number in it?
- What strategy would you use to find the missing number?
- What is the missing number? Explain how you worked it out.
- How can you check your answer?

Ideas for differentiation

Support: Provide these learners with equipment to support their calculations (for example number lines or place value blocks).

Extension: Ask these learners to write their own missing number equations, with totals greater than 200.

Missing number equations

1.

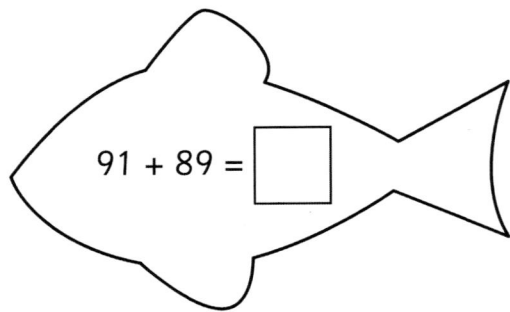

$91 + 89 = \boxed{}$

2.

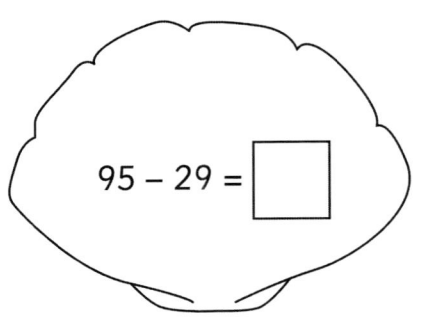

$95 - 29 = \boxed{}$

3.

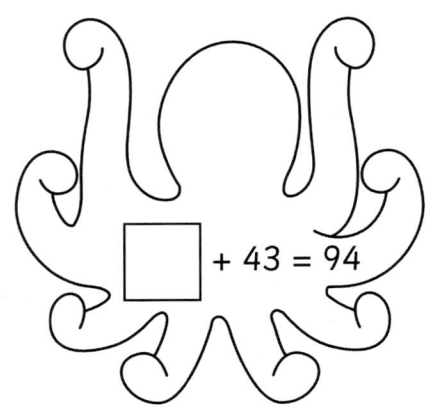

$\boxed{} + 43 = 94$

4.

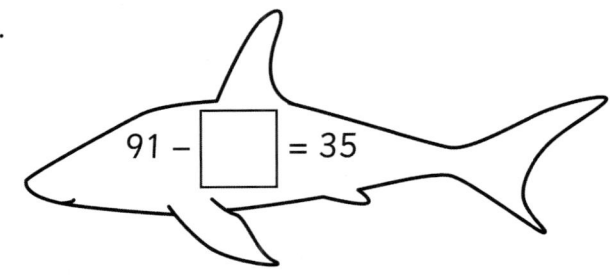

$91 - \boxed{} = 35$

5.

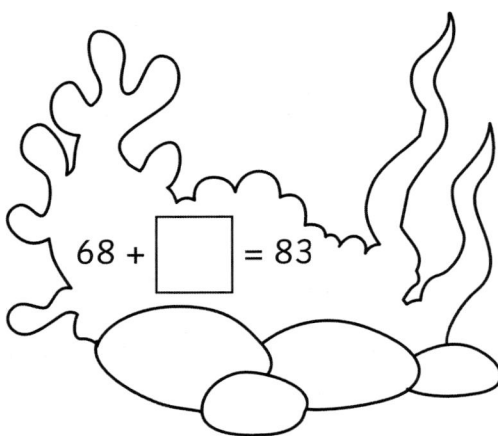

$68 + \boxed{} = 83$

6.

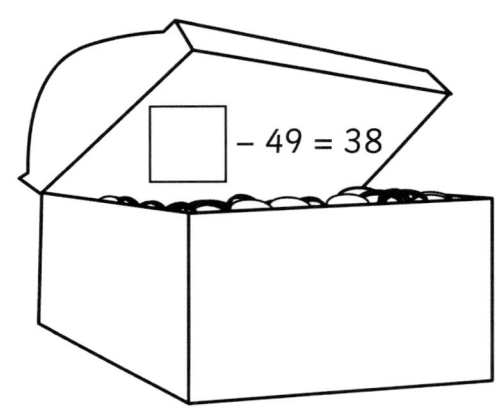

$\boxed{} - 49 = 38$

7.

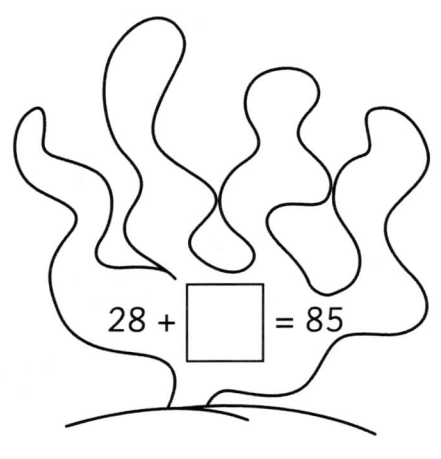

$28 + \boxed{} = 85$

8.

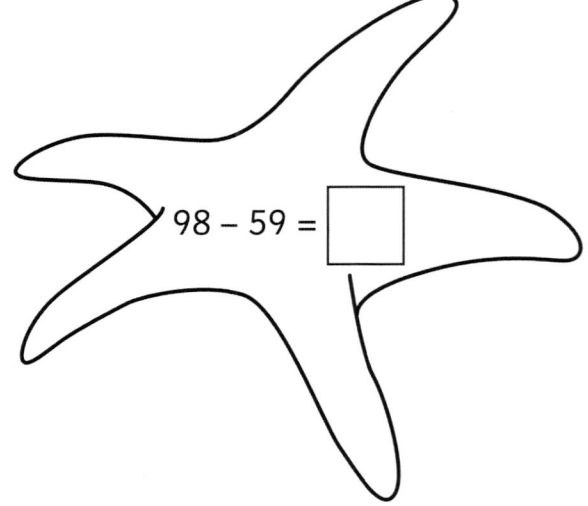

$98 - 59 = \boxed{}$

 Cambridge Primary: Ready to Go Lessons for Maths Stage 3 © Hodder & Stoughton Ltd 2013

Addition and subtraction strategies 4

Learning objectives

- Add three-digit and two-digit numbers using notes to support. (3Nc15)
- Add / subtract single-digit numbers to / from three-digit numbers. (3Nc17)
- Explore and solve number problems and puzzles. (3Ps3)
- Explain methods and reasoning orally, including initial thoughts about possible answers to a problem. (3Ps9)

Resources

Large number cards: set 1 (two lots of whole numbers from 1 to 9), set 2 (two lots of multiples of 10 from 10 to 90) and set 3 (two lots of multiples of 100 from 100 to 900); photocopiable page 140; squared paper.

Starter

- Shuffle each set of number cards separately.
- Hold up two number cards from set 1 (for example 6, 8), asking the learners to call out the total (for example 14). Continue holding up two cards at a time until all the number cards in set 1 have been used. Keep the pace brisk.
- Repeat for the number cards in set 2.
- Repeat for the number cards in set 3.

Main activities

- Write on the board an addition involving at least one three-digit number (for example 75 + 238). Ask the learners to find the answer and record it in a number sentence. Ask the learners to explain how they did the calculation (for example sketching a number line and counting on from the larger number, adding 70 mentally and then adding another 5, partitioning into HTU, then adding and recombining).
- Demonstrate the sorts of informal jottings the learners might want to use to help them keep track of the calculation.

- For example if using the partitioning and recombining method you might record:

$75 = 70 + 5$
$238 = 200 + 30 + 8$
$200 + 70 + 30 + 5 + 8$ $= 200 + 100 + 13$ $= 313$

- Repeat for a couple more additions involving three-digit numbers.
- Display photocopiable page 140 and explain how a cross number puzzle works. (It's just like a crossword puzzle except the answers are numbers instead of words.) Work with the learners to solve a couple of the clues in the puzzle.
- Give each learner photocopiable page 140. Ask the learners to complete the puzzle by working out the answers to the clues.

Plenary

- Write on the board: □□□ + □□□ = 777.
- Ask the learners to complete the number sentence using just the digits 1 to 6, once each. Ask: *Is there more than one solution?* (Yes, there are many, for example 624 + 153!)

Success criteria

Ask the learners:

- How did you work out the answer to this clue?
- How could you check that your answer is correct?
- What clue could you write that would give an answer of 321?

Ideas for differentiation

Support: Give these learners a smaller cross number puzzle with answers below 100.

Extension: Challenge these learners to devise their own cross number puzzle on squared paper, and swap with a friend.

Name: _____

Cross number puzzle

1		2		3		4
		5				
6	7			8	9	
10		11		12		13
		14				
15				16		

Across

1. 130 + 164
3. 144 + 5
5. 70 + 283
6. 65 + 710
8. 139 + 142
10. 712 + 123
12. 68 + 108
14. 193 + 52
15. 498 + 3
16. 235 + 365

Down

1. 180 + 67
2. 232 + 203
3. 90 + 42
4. 955 + 6
7. 399 + 384
9. 77 + 750
10. 411 + 414
11. 513 + 8
12. 73 + 83
13. 250 + 440

Doubling and halving 3

Learning objectives

- Work out quickly the doubles of numbers 1 to 20 and derive the related halves. (3Nc6)
- Work out quickly the doubles of multiples of 5 (<100) and derive the related halves. (3Nc7)
- Work out quickly the doubles of multiples of 50 to 500. (3Nc8)
- Identify simple relationships between numbers. (3Ps6)

Resources

A set of large cards made from photocopiable page 32; sets of cards made from photocopiable page 142.

Starter

- Revise doubling whole numbers 1 to 20 and the related halving facts, and doubling multiples of 10 to 100 and the related halving facts.
- Shuffle a set of large doubling and halving cards made from photocopiable page 32. Hold up the cards one at a time, asking the learners to call out the answer. Keep the pace brisk.

Main activities

- On the board, write the multiples of 5 to 50 in a column. Ask the learners to double each number, describing any patterns in the numbers (for example the doubled numbers all end in zero, are all multiples of 10, are all even, make the 10 times table, their tens digits go up by one each time, and so on).
- Next to the multiples of 5, write the multiples of 50 to 500. Ask the learners to double each number and describe any patterns in the numbers, including the link between the doubles of multiples of 5 and the doubles of multiples of 50.

- Organise the learners into groups of five. Give each group a set of cards made by enlarging photocopiable page 142 onto A3 card (or larger) and cutting out. Ask the learners to shuffle the cards and deal eight cards to each player. Players should take it in turns to place a card from their hand face up on the table (for example half of 600). The player who has the partner card (for example 300) must play it. This player should then choose another card from their hand to play. The game is over when one player has no cards left in their hand. That player is the winner.

Plenary

- On the board, write the multiples of 500 to 5000. Ask the learners to double each number and describe any patterns in the numbers, including the link between doubling multiples of 500 and doubling multiples of 50 and 5.

Success criteria

Ask the learners:

- What's half of 900?
- What's double 45?
- What's half of 70?
- Write a doubling fact you know and the matching halving fact.
- How are the double of 15 and the double of 150 related?

Ideas for differentiation

Support: In the first two main activities, organise these learners to work in a pair with a partner who is more confident.

Extension: In the first two main activities, challenge these learners to continue doubling multiples of 5 past 50 and multiples of 50 past 500.

Doubling and halving cards 2

half of 10	5	double 5	10
half of 30	15	double 10	20
half of 50	25	double 15	30
half of 70	35	double 20	40
half of 90	45	double 25	50
half of 200	100	double 100	200
half of 600	300	double 200	400
half of 1000	500	double 300	600
double 350	700	double 400	800
double 450	900	double 500	1000

Multiplication strategies

- Know multiplication / division facts for 2×, 3×, 5× and 10× tables. (3Nc3)
- Multiply teens numbers by 3 and 5. (3Nc22)
- Check multiplication by reversing the order, for example checking that 6 × 4 = 24 by doing 4 × 6. (3Pt6)

Beanbag; laminated multiplication grids made from photocopiable page 144; dry wipe pens; cloths; plenty of sets of 11 to 19 number cards; six-sided dice labelled 3, 3, 3, 5, 5, 5; calculators; standard six-sided dice.

Starter

- Throw a beanbag to a learner while saying a multiplication in the 3 or 5 times table, for example: *Seven times three.* The learner must throw the beanbag back, saying the product (answer), for example: 'Twenty-one'.
- Throw the beanbag to another learner. Encourage speed.

Main activities

- On the board write 16 × 3. Ask the learners to work out the answer, and explain the strategies they used.
- On the board draw a multiplication grid like those on photocopiable page 144. Demonstrate how to use it to multiply 16 by 3.
  - First partition 16 into 10 + 6. Explain that because 16 = 10 + 6, you can work out 16 × 3 by working out 10 × 3 and 6 × 3 and adding the answers together.
  - Write 10 and 6 in the top row of the grid and 3 below the multiplication sign.
  - Multiply 10 by 3 and write the product in the grid.
  - Multiply 6 by 3 and write the product in the grid.
  - Add the two products together, and write the answer in the 'total' column.

×	10	6	TOTAL
3	30	18	48

- Model checking the result of each multiplication by reversing the order (for example checking 3 × 10 by working out 10 × 3).
- Give each learner photocopiable page 144 (or a laminated card multiplication grid made using the page, and a dry wipe pen and a cloth). Work through a few calculations together (multiplying teens numbers by 3 or 5).
- Organise the learners into ability groups of four. Give each group a set of 11 to 19 number cards and a six-sided dice labelled 3, 3, 3, 5, 5, 5. Ask the learners to generate a teens number using the cards, and generate 3 or 5 by throwing the dice. Players should use their multiplication grid to perform the calculation, and then place their grid face down in a single pile. The first correct answer gets 4 points, the second gets 3, the third 2, and the fourth 1. The winner is the player with the most points when the time is up.

Plenary

- On the board write 36 × 5. Ask the learners to do the calculation on the multiplication grid. Discuss mental methods of multiplying 30 by 5. Practise multiplying other numbers between 20 and 50 by 3 or 5.

Ask the learners:

- Choose a number between 11 and 19. Multiply it by 3 or 5.
- Explain how you worked out the answer.
- How can you check that you've got the right answer?

Support: In the final main activity, give groups of these learners a calculator for checking answers.

Extension: In the final main activity, give groups of these learners a standard six-sided dice. If they throw a 1 they should roll again.

Multiplying a two-digit number by a one-digit number

×			Total

×			Total

×			Total

×			Total

Division strategies

- Begin to divide two-digit numbers just beyond 10× tables, e.g. 60 ÷ 5, 33 ÷ 3. (3Nc23)
- Understand that division can leave a remainder (initially as 'some left over'). (3Nc24)
- Understand the relationship between multiplication and division and write connected facts. (3Nc26)
- Make sense of and solve word problems and begin to represent them. (3Pt3)
- Check a division using multiplication, e.g. check 12 ÷ 4 = 3 by doing 4 × 3. (3Pt7)

Pencils; plain paper; photocopiable page 146; multiplication grids made from photocopiable page 92.

Starter

- Organise the learners into pairs and give each pair paper and pencils. On the board write a division from the 2, 3, 4, 5, 6, 9 or 10 times table (no remainders), for example 20 ÷ 5.
- Ask pairs to write down the answer. Confirm the correct answer. Ask a pair who has written the right answer to explain how they worked it out. Encourage using knowledge of multiplication facts to work out division facts.

Main activities

- Write 70 ÷ 5 on the board, asking the learners to work out the answer. Ask them to describe the strategies they used.
- Model using knowledge of multiplication facts to carry out a remainderless division that goes beyond the times tables, for example 52 ÷ 4: 52 = 40 + 12; 40 ÷ 4 = 10 and 12 ÷ 4 = 3; 10 + 3 = 13 so 52 ÷ 4 = 13. Model how to check answers using multiplication and addition.

- Display a copy of photocopiable page 146. Read one of the problems aloud. Ask: *What calculation do you need to do in order to solve this problem?* Write the calculation on the board. Give pairs time to perform the calculation and then ask them to explain their strategies. Repeat for a second problem.
- Give each pair photocopiable page 146, asking them to work through the rest of the problems.

Plenary

- On the board write a division that extends a little way past the times tables and also has a remainder (for example 133 ÷ 10; 76 ÷ 6; 43 ÷ 3). Ask the learners to calculate the answer, and then discuss strategies.

Ask the learners:

- What calculation did you need to do in order to solve this problem?
- What is the answer to the problem? How did you work it out?
- How could you check your answer?

Support: Group these learners together and work through a couple more problems with them. Also provide them with multiplication grids made from photocopiable page 92.

Extension: These learners should work individually rather than with a partner. Early finishers could devise similar problems to give to a friend.

Name: _____

Division problems

1. 26 crayons are divided equally between two boxes.
   How many crayons are there in each box?

   _____

2. Class 4 has 36 children in it.
   The teacher divides the class into three equal groups.
   How many children are there in each group?

   _____

3. Nina's father is 42. Nina is exactly one third her dad's age.
   How old is Nina?

   _____

4. A square is 48 cm around the outside. What is the length of each side?

   _____

5. 60 biscuits are shared equally among four plates.
   How many biscuits are there on each plate?

   _____

6. Six identical tables have a total mass of 66 kg.
   What is the mass of one table?

   _____

7. There are 126 pages in a book.
   The book is divided into nine equal chapters.
   How many pages are there in each chapter?

   _____

Cambridge Primary: Ready to Go Lessons for Maths Stage 3 © Hodder & Stoughton Ltd 2013

Fractions 1

- Find half of odd and even numbers to 40, using notation such as $13\frac{1}{2}$. (3Nn14)
- Make up a number story to go with a calculation. (3Ps1)

0 to 9 number fans (see photocopiable page 24); large number cards with even numbers from 10 to 40 (but not 22); five apples; knife; cards made from photocopiable page 148.

Starter

- Organise the learners into pairs, giving each pair a 0 to 9 number fan.
- Shuffle the set of large number cards with even numbers from 10 to 40. (There is no 22 because half of 22 cannot be made on the number fans.)
- Holding up one card at a time, ask: *What's half of this number?*
- Ask the learners to make the answer on their number fan. Confirm the correct answers. Challenge the learners by keeping the pace brisk.

Main activities

- Demonstrate finding half of an odd number using practical materials (for example finding half of five apples).
- Introduce how to write one half in numerals. Explain what each number in the fraction represents, introducing the terms numerator and denominator.
- Ask the learners to find half of 17 and write the answer in numerals. Discuss strategies used. Repeat for a few more odd numbers.
- Organise the learners into groups of five. Give each group a set of cards made by enlarging photocopiable page 148 onto A3 card (or larger) and cutting out. Ask the learners to shuffle the cards and deal eight cards to each player.

Players should take it in turns to place a card from their hand face up on the table (for example half of 19). The player who has the partner card (for example $9\frac{1}{2}$) must play it. This player should then choose another card from their hand to play. The game is over when one player has no cards left in their hand. That player is the winner.

Plenary

- Ask: *Was the card game based completely on luck, or could you do something to increase your chance of winning? If so, what could you do?*
- On the board write 'Half of 29', asking the learners to work out the answer. Discuss strategies. Write the answer on the board: 'Half of 29 = $14\frac{1}{2}$'.
- Explain that you are going to make up a number story to go with the calculation. Say: *Lee's petrol tank holds 29 litres of petrol when it is full. Right now it is half full. It contains $14\frac{1}{2}$ litres of petrol.* Ask the learners to make up their own number stories to go with the calculation.
- If you have time, repeat the activities in the previous two bullet points for other halving calculations.

Ask the learners:

- What is half of 28?
- Explain how you worked out the answer.
- What is half of 37?
- Explain how you worked out the answer.
- Tell me what you know about this number: $\frac{1}{2}$

Support: Group these learners together for the final main activity, and work with them, supporting them in their calculations.

Extension: Challenge these learners to perform halving calculations entirely mentally, without making any jottings.

Halving odd numbers cards

half of 1	$\frac{1}{2}$	half of 21	$10\frac{1}{2}$
half of 3	$1\frac{1}{2}$	half of 23	$11\frac{1}{2}$
half of 5	$2\frac{1}{2}$	half of 25	$12\frac{1}{2}$
half of 7	$3\frac{1}{2}$	half of 27	$13\frac{1}{2}$
half of 9	$4\frac{1}{2}$	half of 29	$14\frac{1}{2}$
half of 11	$5\frac{1}{2}$	half of 31	$15\frac{1}{2}$
half of 13	$6\frac{1}{2}$	half of 33	$16\frac{1}{2}$
half of 15	$7\frac{1}{2}$	half of 35	$17\frac{1}{2}$
half of 17	$8\frac{1}{2}$	half of 37	$18\frac{1}{2}$
half of 19	$9\frac{1}{2}$	half of 39	$19\frac{1}{2}$

Fractions 2

Learning objectives

- Understand and use fraction notation recognising that fractions are several parts of one whole, e.g. $\frac{3}{4}$ is three quarters and $\frac{2}{3}$ is two thirds. (3Nn15)
- Recognise equivalence between $\frac{1}{2}$, $\frac{2}{4}$, $\frac{4}{8}$ and $\frac{5}{10}$ using diagrams. (3Nn16)

Resources

Follow-me cards from photocopiable page 84; timer; cards made from photocopiable page 150; photocopiable page 151; paper and pencils.

Starter

- Practise quick-fire addition and subtraction facts by playing a follow-me card game using cards made by enlarging photocopiable page 84 onto A3 card and cutting out. Find and keep the START card. Give out the rest of the cards. (Some learners may need to share one card between two.) Read the START card aloud. The learner who has the answer to that calculation written on their card should read their card aloud, and so on, until the END card is reached.
- Play a few more rounds, redistributing the cards between rounds. Time the activity, and challenge the class to beat their previous best time.

Main activities

- Display the fractions chart on photocopiable page 151. Name each row in the chart (one whole, halves, thirds, ... tenths). Explain how to say and write a fraction made up of more than one block (for example two-thirds = $\frac{2}{3}$, three-quarters = $\frac{3}{4}$). Explain that the denominator names the fraction (for example halves, fifths, tenths) and the numerator gives the number of parts.
- Write these fractions on the board: $\frac{1}{2}$, $\frac{2}{4}$. Ask the learners to find these fractions on the chart and say what they notice about them. (They are equal.) Ask the learners to name any other fractions that are equal to $\frac{1}{2}$, and then any other pairs of fractions that are equal to each other.

- Enlarge photocopiable page 150 onto A3 card and cut out to make fraction match cards. Display one of the fraction diagram cards. Ask: *How many equal parts is the shape divided into?* (For example six.) *How many of those equal parts are shaded?* (For example four.) *What fraction of the shape is shaded?* (For example four-sixths.) Write this fraction on the board in both words and numerals. Repeat for other fraction diagram cards.
- Organise the learners into ability groups of four or six. Give each group a set of fraction match cards. Within the group, the learners should work in pairs to match each fraction diagram with the fraction written in words and in numerals. The pair who wins the race gets a point. The winning pair in the group is the first pair to three points.

Plenary

- Hand out paper and pencils. Call out a fraction, asking the learners to draw a sketch diagram of the fraction, and write the fraction in numerals.

Success criteria

Ask the learners:

- (Pointing to a fraction diagram:) What is this fraction?
- Write the fraction in numerals.
- Explain what each number in the fraction means.
- Name a fraction that is equal to one half. Explain how you know it is equal.

Ideas for differentiation

Support: Give these learners fewer cards to match in the final main activity. Do this by removing cards from the standard set.

Extension: Ask these learners to make their own sets of fraction match cards.

Fraction match cards

	two-thirds	$\dfrac{2}{3}$
	five-eighths	$\dfrac{5}{8}$
	six-tenths	$\dfrac{6}{10}$
	four-sixths	$\dfrac{4}{6}$
	three-fifths	$\dfrac{3}{5}$
	three-quarters	$\dfrac{3}{4}$

Fractions chart

1									

| $\frac{1}{2}$ | | | | | $\frac{1}{2}$ | | | | |

| $\frac{1}{3}$ | | | $\frac{1}{3}$ | | | $\frac{1}{3}$ | | | |

| $\frac{1}{4}$ | | $\frac{1}{4}$ | | $\frac{1}{4}$ | | $\frac{1}{4}$ | | | |

$\frac{1}{5}$	$\frac{1}{5}$	$\frac{1}{5}$	$\frac{1}{5}$	$\frac{1}{5}$

$\frac{1}{6}$	$\frac{1}{6}$	$\frac{1}{6}$	$\frac{1}{6}$	$\frac{1}{6}$	$\frac{1}{6}$

$\frac{1}{8}$	$\frac{1}{8}$	$\frac{1}{8}$	$\frac{1}{8}$	$\frac{1}{8}$	$\frac{1}{8}$	$\frac{1}{8}$	$\frac{1}{8}$

$\frac{1}{10}$	$\frac{1}{10}$	$\frac{1}{10}$	$\frac{1}{10}$	$\frac{1}{10}$	$\frac{1}{10}$	$\frac{1}{10}$	$\frac{1}{10}$	$\frac{1}{10}$	$\frac{1}{10}$

Fractions 3

Learning objectives

- Recognise simple mixed fractions, e.g. $1\frac{1}{2}$ and $2\frac{1}{4}$. (3Nn17)
- Order simple or mixed fractions on a number line, e.g. using the knowledge that $\frac{1}{2}$ comes half way between $\frac{1}{4}$ and $\frac{3}{4}$, and that $1\frac{1}{2}$ comes half way between 1 and 2. (3Nn18)

Resources

Ten identical pictures of half an object (e.g. ten half oranges); blank number line with 20 divisions; cards made from photocopiable page 153; squared paper.

Starter

- Laying down the pictures of half an object one by one, count in halves: *A half, one, one and a half, two, two and a half … five.*
- Draw a number line on the board with ten divisions. Label it in halves from 0 to 5 (0, $\frac{1}{2}$, 1, $1\frac{1}{2}$, 2, ... $4\frac{1}{2}$, 5).
- Using the number line, count on and back in halves.

Main activities

- Display the blank number line with 20 divisions. Label the left end 0, the right end 5, and every fourth division with a whole number. Ask the learners to help you write in the half numbers. Next, point to the division halfway between 0 and $\frac{1}{2}$, asking: *What's this number?* Establish that the smallest divisions are quarters. Label $\frac{1}{4}$ and $\frac{3}{4}$ on the number line. Ask volunteers to label the rest of the quarter divisions.
- Display six cards from the large set of cards made from photocopiable page 153. Model finding each number on the line, and ordering the cards from smallest to largest.

- Ask the learners to make their own fraction number line by copying the number line from the board onto squared paper. Organise the learners into pairs and give each pair a set of cards made by enlarging photocopiable page 153 onto A3 card and cutting out. Ask them to turn over six cards, arrange the cards in order from smallest to largest, and then write down this ordered list of numbers.

Plenary

- Challenge the learners to order a set of six simple and mixed fractions (halves and quarters only) up to 10. Discuss strategies used.
- Challenge the learners to sketch a number line from 0 to 5 marked off in thirds.

Success criteria

Ask the learners:

- (Pointing to a mixed fraction written in figures:) What number is this?
- What number comes halfway between 5 and 6?
- Can you draw a number line from 0 to 5 marked off in halves?
- Draw four cards. Read the number on each card aloud. Put the numbers in order from smallest to largest.

Ideas for differentiation

Support: Support these learners in the final main activity by pairing them with a more confident partner.

Extension: In the final main activity challenge these learners to order the numbers without referring to the number line, using the number line as a checking tool only.

Fraction cards

$\frac{1}{4}$	$\frac{1}{2}$	$\frac{3}{4}$	1
$1\frac{1}{4}$	$1\frac{1}{2}$	$1\frac{3}{4}$	2
$2\frac{1}{4}$	$2\frac{1}{2}$	$2\frac{3}{4}$	3
$3\frac{1}{4}$	$3\frac{1}{2}$	$3\frac{3}{4}$	4
$4\frac{1}{4}$	$4\frac{1}{2}$	$4\frac{3}{4}$	5

Fractions 4

- Begin to relate finding fractions to division. (3Nn19)
- Find halves, thirds, quarters and tenths of shapes and numbers (whole number answers). (3Nn20)
- Begin to know the 4× table. (3Nc4)
- Multiply single-digit numbers and divide two-digit numbers by 2, 3, 4, 5, 6, 9 and 10. (3Nc21)

Times table charts for the 2, 3, 4 and 10 times tables (including some tabletop copies); photocopiable pages 155 and 156; square grid for display; squared paper.

Starter

- Display the 2, 3, 4 and 10 times tables charts. Practise chanting each table.
- On the board write a division related to the 2, 3, 4 or 10 times table (for example $28 \div 4$). Ask the learners to give the answer, linking it to the relevant multiplication (for example $28 \div 4 = 7$ because $7 \times 4 = 28$).

Main activities

- Enlarge photocopiable page 155 onto A3 card or paper and cut out. Display the rectangle and ask a learner to find half of the shape by folding. Compare the number of squares in the whole shape and in half the shape, explaining that finding a half is equivalent to dividing by 2. Ask a learner to fold the shape in half again. Compare the number of squares in the whole shape and in the quarter shape. Explain that finding a quarter is equivalent to dividing by 4. Repeat with the rectangle from photocopiable page 156.

- Draw a five by three rectangle on a square grid. Ask: *What number does this rectangle represent?* (15.) Ask a volunteer to shade in a third of the rectangle. Ask: *What's one third of 15?* Explain that finding a third is equivalent to dividing by 3. Repeat for finding one tenth (for example of 40 using a four by ten rectangle).

- Model finding unit fractions of numbers using division facts only (for example $\frac{1}{3}$ of 24 = 8 because $24 \div 3 = 8$). Give the learners similar questions to answer, finding $\frac{1}{2}$, $\frac{1}{3}$, $\frac{1}{4}$ and $\frac{1}{10}$ of numbers (whole number answers only).

Plenary

- Take in and discuss the learners' answers to the fraction calculations.
- Ask: *How could you check an answer?* (For example by drawing a rectangle on a squared grid; by using counters; by using a calculator to divide, and so on.)

Ask the learners:

- Explain how to find a quarter of a shape.
- Finding a quarter of a number is the same as dividing the number by what?
- What number do I need to divide 80 by to find one tenth of it?
- What is one third of 12?

Support: In the final main activity, give these learners copies of the times table charts for the 2, 3, 4 and 10 times tables.

Extension: Challenge these learners to write more fractions calculations for each other based on the 2, 3, 4 and 10 times tables.

Folding fractions 1

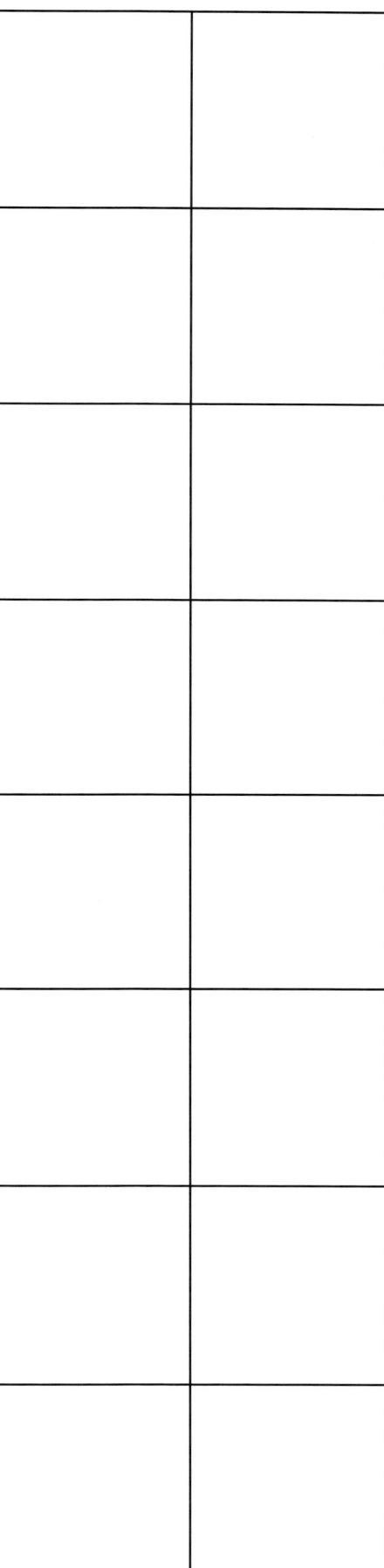

Folding fractions 2

Cambridge Primary: Ready to Go Lessons for Maths Stage 3 © Hodder & Stoughton Ltd 2013

Unit assessment

- What number is 30 less than 726?
- What do you have to add to 42 to make 100?
- How could you check that 36 ÷ 4 = 9 is correct?

- Name three fractions that are equivalent to one half. Explain how you worked it out.
- Finding one third of a number is the same as dividing by what?

Summative assessment activities

Observe the learners while they play these games. You will quickly be able to identify those who appear to be confident and those who may need additional support.

Function machine game

This game assesses the learners' ability to find multiples of 10 and 100 more or less than three-digit numbers.

You will need:

Laminated copy of photocopiable page 130; dry wipe pen; 0 to 9 ten-sided dice; six-sided dice labelled +, +, +, −, −, −; number cards 20, 30, 40, 50, 60, 70, 80, 90, 200, 300; calculators; scrap paper; pencils.

What to do

- Organise the learners into groups of six. The start player should use a ten-sided dice to generate a three-digit number and write it in the 'input' box on photocopiable page 130, then roll the six-sided dice and turn over a number card. They should write the result (for example −40) inside the function machine.

- Players should calculate the output number and write it on pieces of scrap paper with their names on then place these in a single pile face down. The start player should check the answer with a calculator. The first player with the right answer gets 3 points, the second 2 points and everyone else 1 point.

- Repeat until all the players have rolled the dice twice. The player with the most points is the winner.

Fraction match game

This game assesses the learners' ability to understand and use fraction notation.

You will need:

Cards from photocopiable page 150.

What to do

- Organise the learners into groups of five. Each group should choose a caller. The caller should keep the cards showing fractions written in words, shuffle the remaining cards together and deal them to the other four players.

- The caller should read out a 'word fraction' card. Any player with that fraction should place the card face up. The first player to do this wins the word card. The round is over when all the word cards have been won. Each player scores points equal to the number of word cards they have.

- Play another round with a different player acting as caller. The game ends when every player has been caller. The player with the most points wins.

Distribute photocopiable page 158. Ask the learners to read and answer the questions. They should work independently.

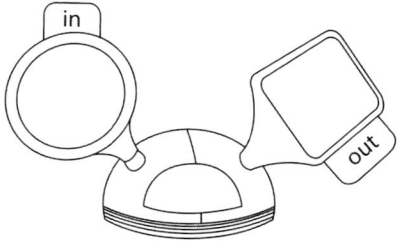

Name: _____

Investigating numbers

1. Double these numbers.

   a) 450 _____

   b) 350 _____

   c) 150 _____

   d) 250 _____

2. What is 15 × 5? _____

3. Eduardo is packing soft-drink cans into boxes of six. He has 72 cans to pack. How many full boxes can he make?

   _____

4. Write a number between each pair of numbers.

   a) 683 and 677 _____

   b) 825 and 829 _____

   c) 404 and 410 _____

   d) 900 and 889 _____

5. Draw a number line and put these numbers on it: 3, $\frac{3}{4}$, $4\frac{3}{4}$, $2\frac{1}{4}$, $3\frac{1}{2}$, $4\frac{1}{4}$, $2\frac{3}{4}$, $1\frac{1}{2}$.

6. Shade in one quarter of this square.

7. Circle one third of the bananas.

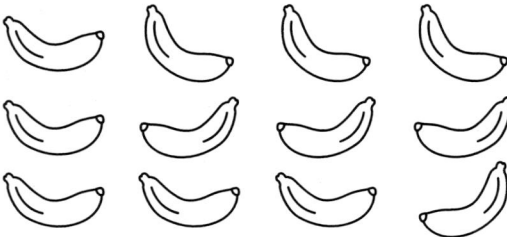

8. What is:

   a) $\frac{1}{2}$ of 16? _____

   b) $\frac{1}{3}$ of 21? _____

   c) $\frac{1}{4}$ of 24? _____

   d) $\frac{1}{10}$ of 90? _____

Cambridge Primary: Ready to Go Lessons for Maths Stage 3 © Hodder & Stoughton Ltd 2013

2D shapes 3

Learning objectives

- Identify, describe and draw regular and irregular 2D shapes including pentagons, hexagons, octagons and semi-circles. (3Gs1)
- Classify 2D shapes according to the number of sides, vertices and right angles. (3Gs2)

Resources

2D shapes for display and for tabletop use (e.g. circles, semi-circles, triangles, squares, rectangles, parallelograms, kites, rhombuses, pentagons, hexagons and octagons); photocopiable page 160; A3 paper; mirrors.

Starter

- Hold up 2D shapes one at a time, asking the learners to name them. When each shape has been named, describe one of its properties, for example: *An equilateral triangle has three equal sides.* Ask the learners to describe the shape's other properties, for example three equal angles.
- Display all the 2D shapes. Describe the properties of one of the shapes without naming it or pointing to it. Ask the learners to point to and name the shape. Repeat the activity, with a learner choosing and describing a shape.

Main activities

- Display a copy of photocopiable page 160. Help the learners to sort the 2D shapes already on display into the first Carroll diagram on photocopiable page 160. Record the groupings by drawing round the shapes. Repeat the activity with the second Carroll diagram on photocopiable page 160. Draw a third Carroll diagram on the board, asking the learners to suggest the labels.

- Organise the learners into pairs, giving each pair several sheets of A3 paper and a selection of 2D shapes. Ask the learners to copy the third Carroll diagram from the board onto A3 paper, and sort their 2D shapes into it, recording the groupings by drawing around each shape.
- Finally, ask the learners to sort the shapes on their table according to their own criteria. They may choose to draw another Carroll diagram, or use a different sorting method (for example a Venn diagram or a simple sorting table).

Plenary

- Ask the learners to share the Carroll diagrams they have drawn with the rest of the class.
- Display a selection of 2D shapes in an unlabelled Carroll diagram. Ask the learners to work out the criteria that you used in order to sort the shapes, and suggest an appropriate label for each box in the diagram.

Success criteria

Ask the learners:

- Draw an irregular pentagon.
- Name this shape. Describe one of its properties.
- Sort these shapes according to their number of sides.
- Can you write in the missing labels on this Carroll diagram?

Ideas for differentiation

Support: Group these learners together for the main activity, and work with them.

Extension: Challenge these learners to draw at least one extra shape (for example a shape that they don't have a template for) in each box of their Carroll diagrams.

Sorting 2D shapes: Carroll diagrams

	at least 1 curved side	no curved sides
4 or more vertices		
fewer than 4 vertices		

	at least 1 right angle	no right angles
at least 2 equal sides		
no equal sides		

2D shapes 4

- Recognise the relationships between different 2D shapes. (3Pt8)

2D shapes for display (such as those used in the previous lesson); cards made from photocopiable page 162.

Starter

- Display two 2D shapes that have at least one property in common. Ask the learners to describe something that is the same about both shapes. Repeat for other pairs of shapes, for example rectangle and kite; regular hexagon and equilateral triangle; isosceles triangle and semi-circle; circle and semi-circle.

Main activities

- Enlarge photocopiable page 162 onto A3 card to make 2D shape cards and cut them out.
- Teach the following game: Call up two volunteers. Shuffle a large pack of 2D shape cards. Deal six cards to each volunteer and display them. Ask the first volunteer to choose a card from their hand to play. To play the card they must describe three of its properties. The next player must play a card that has at least one property in common with the card just played. (The property does not have to be one of those the previous player described.) The player should play a shape, describe the property it has in common with the previous card and describe two more of its properties. If a player does not have a card that has properties in common with the last card played, play passes to the next player. The first player to have no more cards left in their hand is the winner.
- Organise the learners into groups of three to play the game. Give each group a set of 2D shape cards made by enlarging photocopiable page 162 onto A3 card and cutting out.

Plenary

- Ask the learners to identify as many shapes as they can with a given property, for example ask: *Which shapes have no angles at all? Which shapes have two pairs of equal sides? Which shapes are symmetrical?*
- Ask a volunteer to choose and hold up two 2D shapes that have at least one property in common. Ask the rest of the learners to describe the property (or properties) the shapes have in common.

Ask the learners:

- Choose two 2D shapes that have something in common.
- Describe the property or properties the shapes you chose have in common.

Support: In the final main activity, group these learners together, and work with them during the game to prompt their thinking about the similarities between shapes.

Extension: In the final main activity, challenge these learners to describe four properties of each shape.

More 2D shape cards

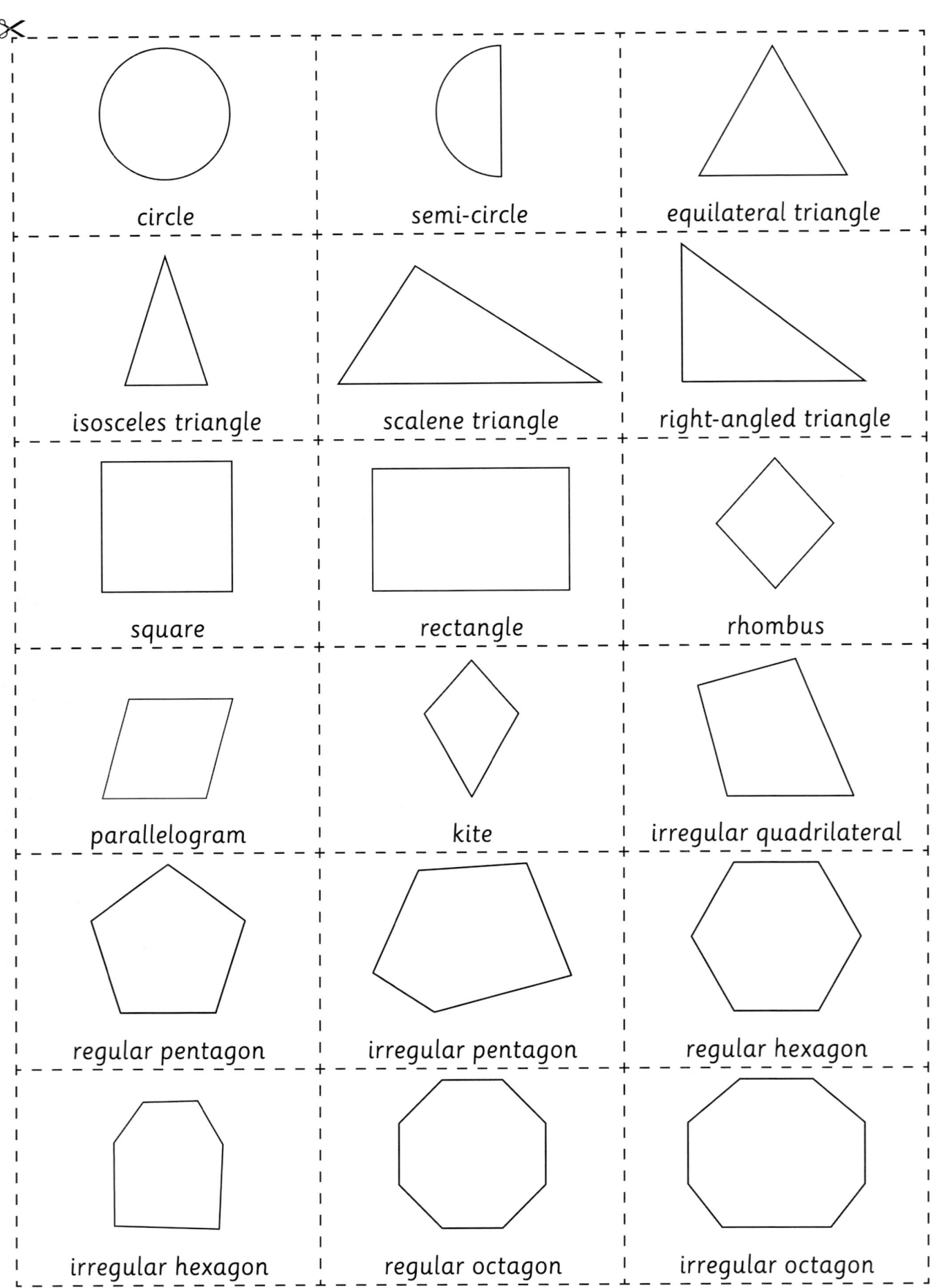

circle	semi-circle	equilateral triangle
isosceles triangle	scalene triangle	right-angled triangle
square	rectangle	rhombus
parallelogram	kite	irregular quadrilateral
regular pentagon	irregular pentagon	regular hexagon
irregular hexagon	regular octagon	irregular octagon

Cambridge Primary: Ready to Go Lessons for Maths Stage 3 © Hodder & Stoughton Ltd 2013

Symmetry 2

Learning objectives

- Draw and complete 2D shapes with reflective symmetry and draw reflections of shapes (mirror line along one side). (3Gs5)
- Identify simple relationships between shapes, e.g. these shapes all have the same number of lines of symmetry. (3Ps7)

Resources

Large and small 2D shapes; mirrors, including a large mirror; pencils; plain paper; photocopiable page 164.

Starter

- Hold up one large 2D shape at a time. Ask the learners to point their thumbs up if the shape is symmetrical and point their thumbs down if the shape is not symmetrical.
- Display all the symmetrical shapes. Ask the learners to work in pairs to sort these shapes into groups according to their number of lines of symmetry.
- Ask pairs to explain how they sorted the shapes. Clarify any misconceptions by using a large mirror to show the lines of symmetry. Discuss the special case of the circle, which has an infinite number of lines of symmetry.

Main activities

- Give each learner paper and a pencil. Display a large 2D shape, drawing a dotted 'mirror line' along one edge. Ask the learners to visualise what the shape would look like reflected in a mirror placed along the dotted line, and to sketch what they 'see'.
- Reveal the correct answer by placing the large mirror along the mirror line. Draw the reflected shape on the other side of the mirror line. Repeat for other shapes, with mirror lines always along one edge of the shape. You could demonstrate the technique of flipping the shape over the dotted line to mimic reflection.

- Give each learner a mirror and give each table plenty of 2D shapes. Ask the learners to repeat the activity at their tables using the 2D shapes and mirrors, recording both their visualisation sketches and the actual reflections.

Plenary

- Display a copy of photocopiable page 164. Ask the learners to visualise and sketch each reflected shape. Draw in each correctly reflected shape on the display copy.

Success criteria

Ask the learners:

- How many lines of symmetry does this shape have?
- (Pointing to shapes sorted according to their lines of symmetry:) What label would you give each of these groups of shapes?
- Sort these shapes according to the number of lines of symmetry they have.
- Sketch what this shape would look like reflected in a mirror line along this side.

Ideas for differentiation

Support: In the final main activity, ask these learners to work in pairs.

Extension: In the final main activity, ask these learners to place some mirror lines so that they touch the shape at a vertex rather than along one side.

Shape reflections 2

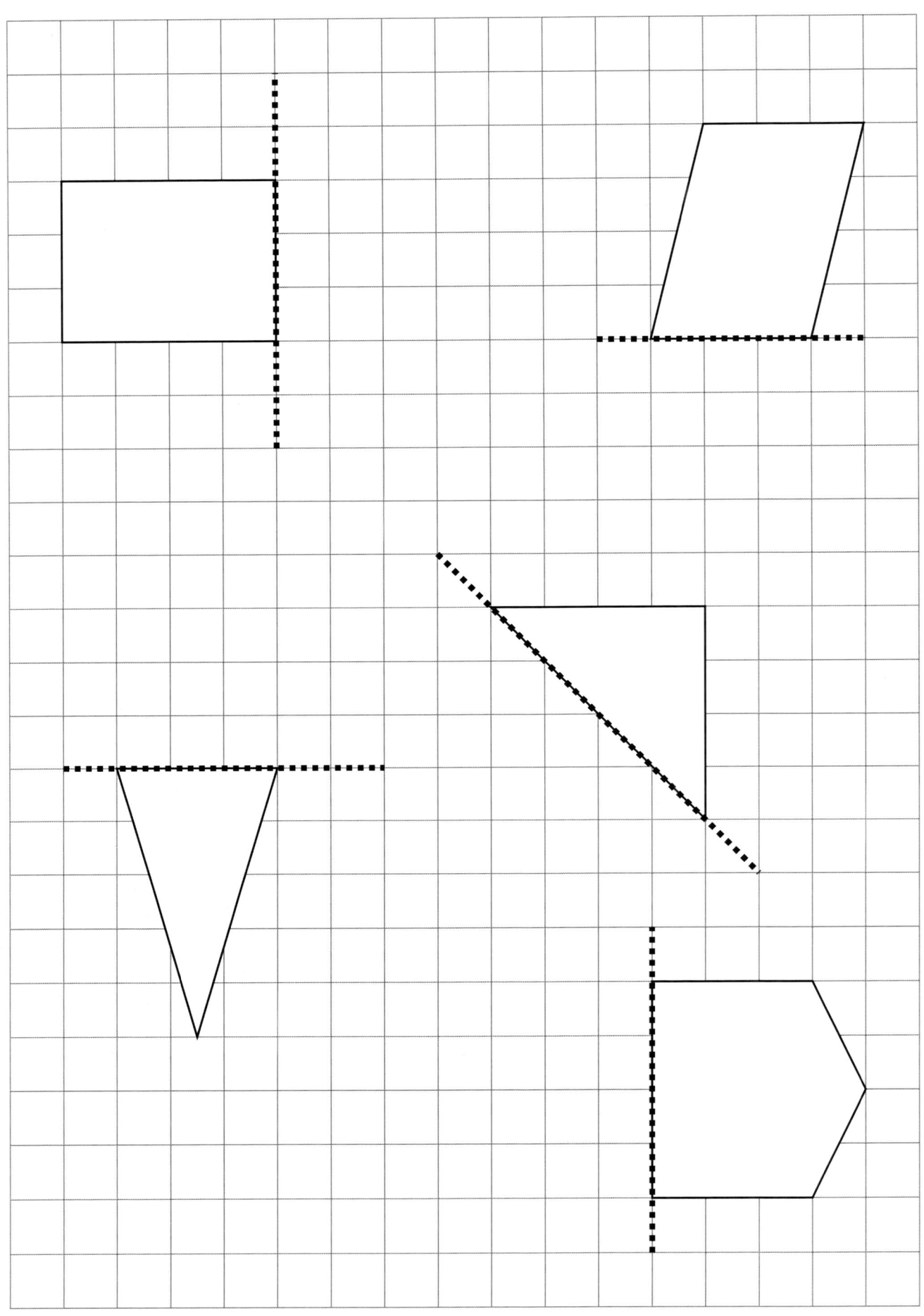

Cambridge Primary: Ready to Go Lessons for Maths Stage 3 © Hodder & Stoughton Ltd 2013

Angles 2

- Identify right angles in 2D shapes. (3Gs8)
- Use a set square to draw right angles. (3Gp3)
- Compare angles with a right angle and recognise that a straight line is equivalent to two right angles. (3Gp4)

Set squares, including a large one; photocopiable page 166; A3 paper; glue sticks; a range of 2D shapes.

Starter

- Ask the learners to stand in a space indoors, all facing the same wall. Call out instructions for turning a certain number of right angles in a particular direction. Call out three or four instructions in a row, for example: *Turn one right angle to the right, turn two right angles to the left, turn three right angles to the right,* and then pause. Ask: *What single instruction could I have given you that would have made you end up facing the same way?* (For example turn two right angles – in either direction.)
- Repeat for other combinations of instructions.

Main activities

- Draw this table on the board:

Less than one right angle	Equal to one right angle	Greater than one right angle but less than two right angles	Equal to two right angles

- Compare a couple of the angles from photocopiable page 166 to a right angle, and place each angle in the correct part of the table.
- Give each learner a copy of photocopiable page 166, a sheet of A3 paper, and a glue stick. Ask the learners to copy the table from the board onto the A3 sheet, and use a set square to measure the angles on photocopiable page 166. They should cut and stick each angle in the correct part of the table.

Plenary

- Ask the learners to describe how they have sorted the angles. Revise the terms 'acute' and 'obtuse', and introduce the term 'straight line angle'.
- Display a range of 2D shapes, asking the learners to identify the right angles in them.

Ask the learners:

- (Pointing to one of the angles from photocopiable page 166 that hasn't been sorted yet:) Which part of the table do you think this angle belongs in? How can you check?
- (Pointing to one of the columns in the table:) Draw an angle to go in this column.

Support: Assist these learners when they are comparing the angles on photocopiable page 166 with the right angle on their set square.

Extension: Challenge these learners to extend the table, adding a column labelled 'more than two right angles', and drawing in their own examples of angles that fit this description.

Angles 2

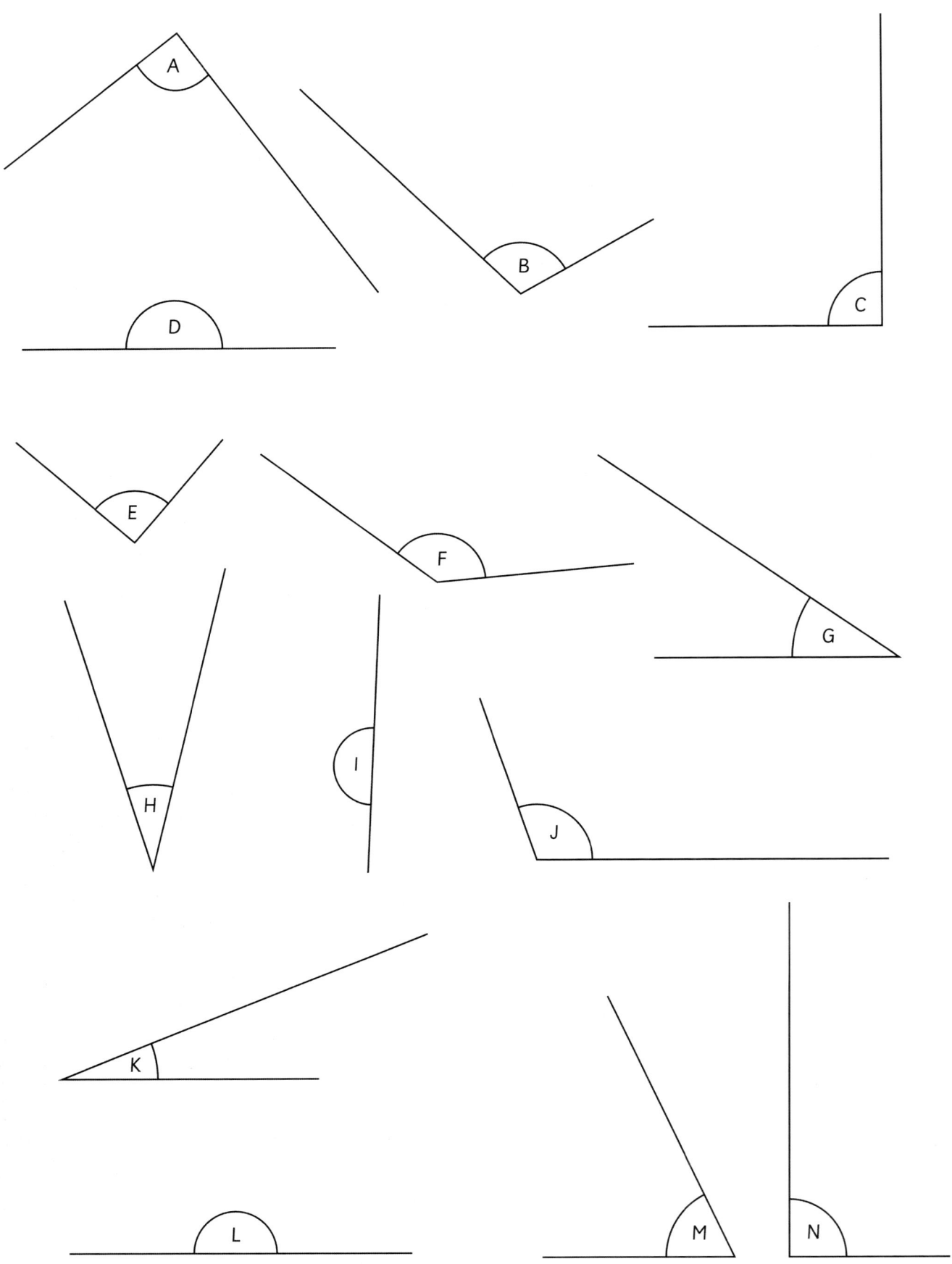

Cambridge Primary: Ready to Go Lessons for Maths Stage 3 © Hodder & Stoughton Ltd 2013

Co-ordinates 2

Learning objectives

- Use the language of position, direction and movement, including clockwise and anti-clockwise. (3Gp1)
- Find and describe the position of a square on a grid of squares where the rows and columns are labelled. (3Gp2)

Resources

Pencils; plain paper; objects to position; photocopiable page 168.

Starter

- Give each learner paper and a pencil. Position a pair of objects and describe the relationship between them, leaving out the word or phrase describing position, for example put a box under a chair, saying: *The box is **something** the chair.* Ask the learners to write the missing word or words. Include a variety of positional vocabulary, including describing rotation as clockwise or anti-clockwise.

Main activities

- Introduce number-only co-ordinates, for example (8, 7). Explain that the first number gives the number across and the second gives the number up. Practise number-only co-ordinates by playing a game of four in a row on a large copy of photocopiable page 168. Divide the class into two teams and call up one volunteer from each team. Team mates must tell the volunteers where to place their mark by calling out co-ordinates. The winning team is the first to make four marks in a row, vertically, horizontally or diagonally.

- Organise the learners into pairs and give each pair photocopiable page 168. Ask the learners to sit opposite their partner and put something between them so that they cannot see each other's grid. Partners should take it in turns to draw a shape in one of the squares on the grid. They should describe the shape and say the co-ordinates of the square. Their partner should draw the same shape in the same square on their grid. Repeat until there are 20 shapes drawn in the grid, then partners can compare grids. They should look the same!

Plenary

- Display a large blank copy of photocopiable page 168. Ask the learners to visualise drawing a dot in the centre of various squares identified by their co-ordinates. Say: *Those dots are the vertices of a shape. What is the shape?* For example you could ask the learners to draw a dot in the centre of each of the following squares: (7, 1), (9, 1) and (8, 3). The learners will find that these dots form the vertices of a triangle.

Success criteria

Ask the learners:

- Can you describe the position of these two objects in relation to each other?
- What are the co-ordinates of that square?
- Point to (3, 6) on the co-ordinate grid.

Ideas for differentiation

Support: Give these learners a visual resource that they can refer to that reminds them in which order to read number-only co-ordinates, such as a pair of arrows in place of the numbers in a co-ordinate pair: ($\rightarrow$, $\uparrow$).

Extension: Challenge these learners to make up their own game using number-only co-ordinates.

Number-only co-ordinates

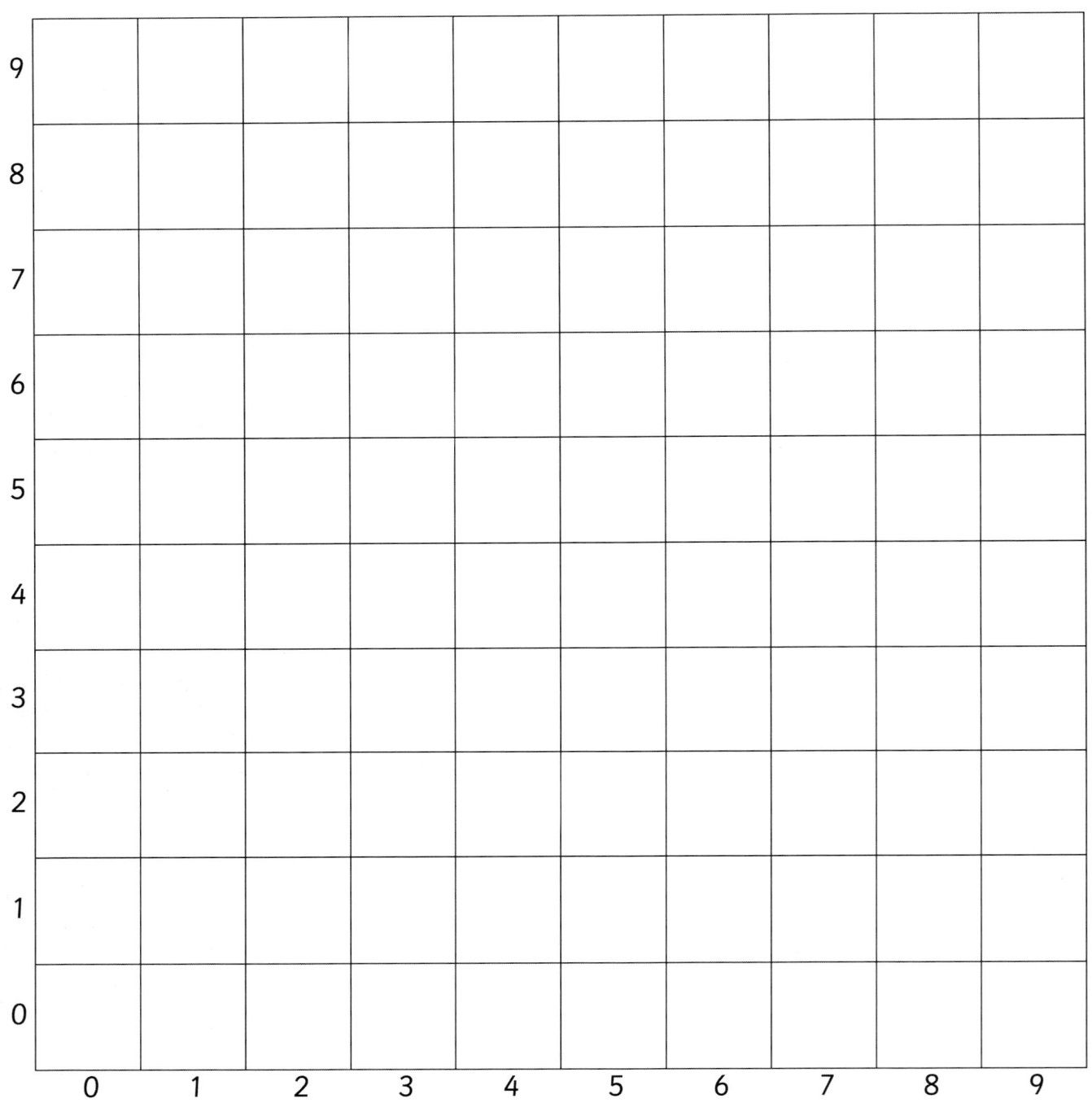

3D shapes 3

- Classify 3D shapes according to the number and shape of faces, number of vertices and edges. (3Gs4)
- Relate 2D shapes and 3D solids to drawings of them. (3Gs6)
- Identify the differences and similarities between different 3D shapes. (3Pt9)

Sets of cards made from photocopiable page 170, including one large set; 3D shapes, such as spheres, hemispheres, cones, cylinders, various pyramids and various prisms, including cubes, cuboids and triangular prisms; A3 paper; pencils.

Starter

- Organise the learners into pairs. Give each pair a small set of 3D shape cards made by enlarging photocopiable page 170 onto A3 card and cutting out. Ask pairs to match each shape card with its correct name card.
- Hold up a model of a 3D shape that is illustrated on photocopiable page 170. Ask pairs to hold up the matching picture card and name card. Repeat for other models of 3D shapes.

Main activities

- Hold up a 3D shape. Ask the following questions: *How many faces has this shape got? What shape are they? How many edges has this shape got? How many vertices has it got? (Vertices are points where edges meet.)* Repeat for several different shapes.
- Ask the learners to sort 3D shapes according to a) the number of faces, b) the shape of faces, c) the number of vertices and d) the number of edges.

- Display a cube and cuboid. Ask: *What is the same about these two shapes?* Write the learners' suggestions in a column labelled 'Similarities'. Ask: *What is different about these two shapes?* Write the learners' suggestions in a column labelled 'Differences'. Encourage the learners to find as many similarities and differences as possible.
- Organise the learners into pairs. Give each table a wide selection of 3D shapes and some A3 paper. Ask the pairs to choose two shapes, and record their similarities and differences.

Plenary

- Ask volunteers to explain the similarities and differences between the pair of shapes they chose to examine. Ask the rest of the learners to describe any other similarities or differences between these shapes.

Ask the learners:

- Choose a 3D shape. Name it. How many faces does it have? How many edges? How many vertices?
- Choose another 3D shape. Compare it with the first shape you chose. What is the same about the two shapes? What is different?

Support: In the final main activity, pair these learners with a partner who is more confident.

Extension: In the final main activity, ask these learners to compare and contrast a second pair of shapes.

3D shape cards

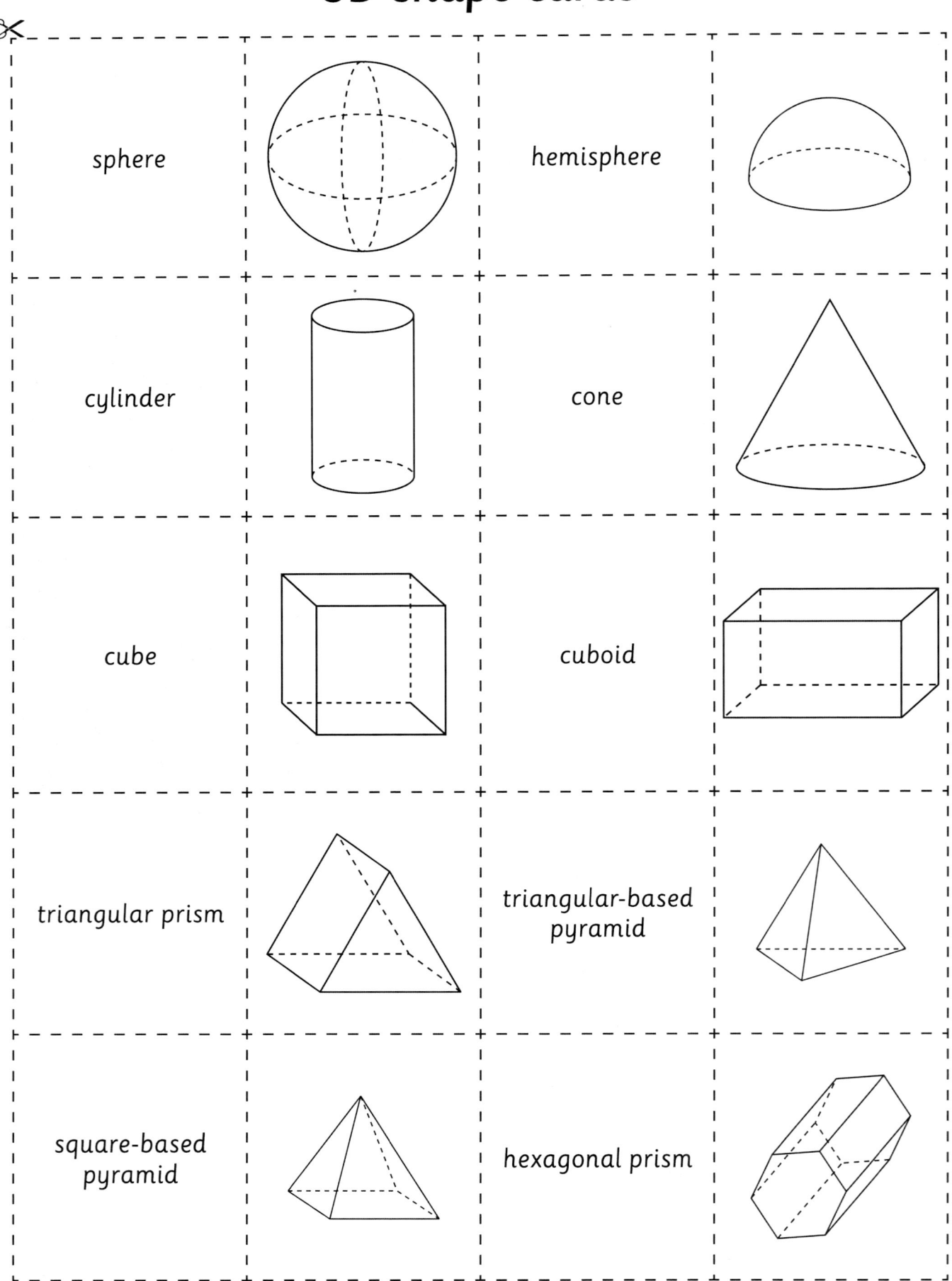

sphere		hemisphere
cylinder		cone
cube		cuboid
triangular prism		triangular-based pyramid
square-based pyramid		hexagonal prism

Cambridge Primary: Ready to Go Lessons for Maths Stage 3 © Hodder & Stoughton Ltd 2013

3D shapes 4

Learning objectives

- Identify, describe and make 3D shapes including pyramids and prisms. (3Gs3)
- Identify 2D and 3D shapes, lines of symmetry and right angles in the environment. (3Gs7)

Resources

Range of 3D shapes, each labelled with a unique number; paper; pencils; photocopiable page 172 or your own classroom shape trail; coloured pencils.

Starter

- Hold up 3D shapes one at a time, asking the learners to call out the name of each shape. Demand specific rather than general names, for example 'square-based pyramid' rather than 'pyramid'.
- Display all the shapes, making sure their numbers are clearly visible. Give each learner paper and a pencil. Describe the properties of one of the shapes, and then ask the learners to write the number of that shape.

Main activities

- Hand out copies of the classroom shape trail (photocopiable page 172 or your own version). Give the learners time, working in pairs, to record their answers.
- Discuss the learners' answers to the questions in the shape trail. Ask the learners whether there are any questions that may have more than one correct answer. (The answers to questions 3 and 4 will depend on the window / pattern chosen, and the answer to 1b is constantly changing, as the hands of the clock turn around.)
- Organise the learners into groups, asking each group to devise their own shape trail around the school and / or playground for others to follow.

Plenary

- Organise time for groups to follow each other's shape trails and report back on the experience.

Success criteria

Ask the learners:

- What shape is this? Is it a prism or a pyramid? What sort of prism / pyramid? Tell me something that all prisms / pyramids have in common.
- Tell me something you found out doing the shape trail.

Ideas for differentiation

Support: In the final main activity, group these learners together and work with them to help them devise their shape trail.

Extension: Challenge these learners to include specific types and numbers of questions on their shape trail (for example at least two questions on each of the following topics: right angles, identifying 2D shapes, identifying 3D shapes, symmetry).

Name: _____

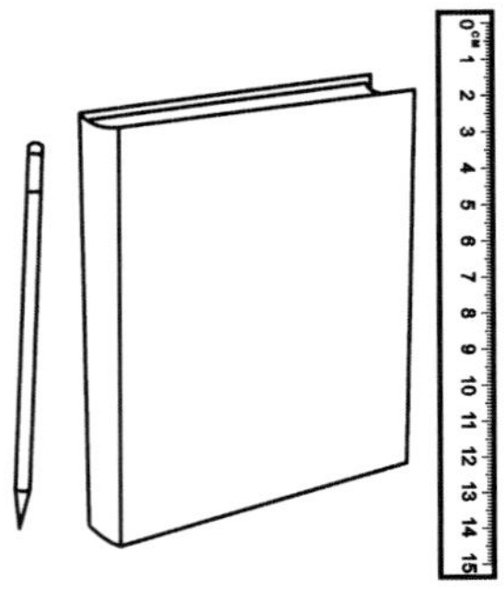

Classroom shape trail

1. Look at the classroom clock.

   a) What shape is the clock face? _____

   b) Look at the angle made by the hands of the clock.
      Circle the right answer:

      less than a right angle   one right angle

      more than one right angle   two right angles

2. Draw three objects in the classroom that are cuboids.

3. Look at one of the windows.

   a) What shape is the window? _____

   b) Draw a sketch of the window.

   c) Use a coloured pencil to mark each line of symmetry on your
      sketch of the window.

4. Look up, look down, look all around. Can you see a pattern anywhere?

   a) Draw a pattern you can see.

   b) Which shapes make up this pattern? _____

   c) How many lines of symmetry does the pattern have? _____

Cambridge Primary: Ready to Go Lessons for Maths Stage 3 © Hodder & Stoughton Ltd 2013

Unit assessment

Questions to ask

- I am thinking of a shape. It has two long sides that are equal and two short sides that are equal. It does not have any right angles. What shape could I be thinking of?
- (Pointing to a selection of 3D shapes:) Which of these shapes are prisms?

- How are all prisms the same?
- (Pointing to a 10 by 10 square grid:) Can you label this grid to make it into a co-ordinate grid?
- (Pointing to a square on the grid labelled in the last question:) What are the co-ordinates of this square?

Summative assessment activities

Observe the learners while they take part in these activities. You will quickly be able to identify those who appear to be confident and those who may need additional support.

Line them up!

This activity assesses the learners' ability to recognise the properties of 2D shapes and see the relationships between them.

You will need:

2D shapes.

What to do

- Organise the learners into table groupings. Give each group a wide selection of 2D shapes.
- Put the shapes in the middle of the table. The aim of the activity is to make a line of 2D shapes around the edge of the table.
- The learners should take it in turns to place a shape on either end of the line. Each new shape must have something in common with the shape it is placed next to. The learner who places the shape must describe this common property.

Make a shape

This activity assesses the learners' ability to name, make and classify 3D shapes, as well as identify similarities and differences between them.

You will need:

3D shapes; art straws; modelling clay; card; scissors; rulers; sticky tape; digital camera; computer; printer; photocopiable page 174.

What to do

- Ask each learner to choose a 3D shape and make an accurate model of it. Explain that they can choose to make the model from modelling clay and art straws, or from card.
- When the learners have finished making their shapes, ask them to take photographs of their models using a digital camera. Print out the photographs. Small images printed onto ordinary printer paper are sufficient.
- Ask the learners to complete photocopiable page 174, sticking the photo of their model onto the page.

Written assessment

Give each learner photocopiable page 168. Ask them to draw an imaginary map on the grid, drawing in and labelling about a dozen landmarks. Each landmark should be completely contained in a single square.

Ask each learner to write half a dozen questions about the landmarks on the map for a friend to answer.

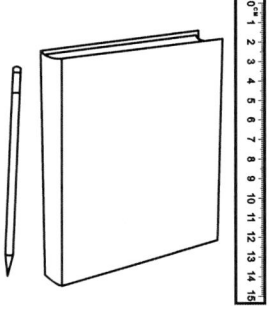

Name: _____

Make a shape

1. Name the 3D shape you have made: _____

2. Draw your shape here or stick on a photo:

3. Answer these questions about your shape:

    a) How many faces does it have? _____

    b) How many edges does it have? _____

    c) How many vertices does it have? _____

4. Now choose another 3D shape to compare your shape to.

    a) What is the name of the other 3D shape? _____

    b) What are the similarities between your shape and the other shape?

        _____

        _____

    c) What are the differences between your shape and the other shape?

        _____

        _____

Cambridge Primary: Ready to Go Lessons for Maths Stage 3 © Hodder & Stoughton Ltd 2013

Unit 3C: Measure and problem solving

Time 7

Learning objectives

- Solve word problems involving measures. (3Ml5)
- Suggest and use suitable units to measure time and know the relationships between them (second, minute, hour, day, week, month, year). (3Mt1)
- Read a calendar and calculate time intervals in weeks or days. (3Mt4)

Resources

Cards made from photocopiable page 104; timer; photocopiable page 176; calculators.

Starter

- Play a follow-me card game, using cards made from photocopiable page 104. Find and keep the START card. Hand out the rest of the cards. (Some learners may need to share one card between two). Read the START card aloud. The learner who has the answer to that question must read their card aloud, and so on, until the END card is reached.
- Play a few more rounds, redistributing the cards between rounds. Time the activity, and challenge the class to beat their previous best time.

Main activities

- Display a copy of photocopiable page 176, which contains problems linked to two pages from a calendar. Work through a couple of the questions together.
- Ask the learners to complete photocopiable page 176, either working individually or in pairs. Challenge early finishers to write extra questions. Discuss answers, and work through some of the extra questions.
- Pose some open-ended questions, asking the learners which units they would use to measure various time intervals, for example: *Which units would you use to measure the time taken by a journey? The time it takes an athlete to run a race? How long you have been alive?*

- Ask: *How long have you been alive?* Ask the learners to calculate and record their current age in: a) years and months, b) days and c) hours. Ensure the learners have access to calculators.

Plenary

- Ask the learners to explain how they worked out each part of the answer in the final main activity. Discuss their answers. Discuss the need for estimation and approximation when dealing with smaller units of time (see 'Extension' below).

Success criteria

Ask the learners:

- Which unit would you use to measure the length of time it takes a plant to grow? Explain your reasoning.
- Suggest an event that you would measure in seconds.
- How many seconds are in a minute / hours are in a day / months are in a year?
- Explain how you worked out the answer to one of the problems on photocopiable page 176.

Ideas for differentiation

Support: For the second main activity, organise these learners to work with a more confident partner. For the final main activity, group these learners together and guide them through the problem.

Extension: For the final main activity, ask these learners to also calculate their age in minutes and in seconds.

Name: _____

Calendar problems

June						
M	T	W	Th	F	Sa	Su
	1	2	3	4	5	6
7	8	9	10	11	12	13
14	15	16	17	18	19	20
21	22	23	24	25	26	27
28	29	30				

July						
M	T	W	Th	F	Sa	Su
			1	2	3	4
5	6	7	8	9	10	11
12	13	14	15	16	17	18
19	20	21	22	23	24	25
26	27	28	29	30	31	

1. What is the date two weeks and six days before Sunday 18th July?

   _____

2. What is the date three weeks and four days after Wednesday 2nd June?

   _____

3. How many weeks and days before Thursday 8th July is Saturday 5th June?

   _____

4. What is the date 15 days before Tuesday 20th July?

   _____

5. What is the date 17 days before Saturday 3rd July?

   _____

6. What is the date nine days after Monday 28th June?

   _____

7. How many days after Friday 18th June is Saturday 3rd July?

   _____

9. Hana is going on holiday from Saturday 26th June to Friday 9th July. How long is that:

   a) in weeks and days? _____

   b) in days? _____

Cambridge Primary: Ready to Go Lessons for Maths Stage 3 © Hodder & Stoughton Ltd 2013

Time 8

Learning objectives

- Read the time on analogue and digital clocks, to the nearest 5 minutes on an analogue clock and to the nearest minute on a digital clock. (3Mt2)

Resources

Pencils; plain paper; sets of cards made from photocopiable page 178 on which you have drawn 16 different times (digital times to the nearest minute and analogue times to the nearest 5 minutes); sets of blank cards made from photocopiable page 178; small cardboard clock faces with moveable hands.

Starter

- Give each learner paper and a pencil. Write one of these multiples of 5 on the board: 30, 35, 40, 45, 50 or 55. Ask: *How many more to make sixty?* Ask the learners to write the answer. Discuss strategies used, for example counting up in fives. Repeat a few times.

- Extend to numbers that are not multiples of 5, for example 33, 37, 42, 49, 54 or 58. Discuss strategies used, for example compensating from multiples of 5, or using knowledge of complements to 10.

Main activities

- Organise the learners into pairs and give each pair a set of time cards made by enlarging photocopiable page 178 onto A3 card and cutting out. Ask the learners to order the times from earliest to latest.

- Discuss the correct order, encouraging the learners to express times in different ways. Draw on the work done in the starter activity to express digital times past the half hour as times to the next hour.

- Hold up one of the time cards. Ask: *What did you do / will you be doing at this time today?*

- Ask the learners to discuss with their partners the activities that they have done / will be doing at those times today.

- Give each pair a set of blank cards made from photocopiable page 178, asking them to make their own set of time cards. When they have finished, ask them to swap with another pair and order each other's cards.

Plenary

- Give each learner a clock face with moveable hands. Call out a time (including a.m. or p.m.), asking the learners to make it on their clock faces.

- For each time made in this way, ask the learners to suggest an activity you might do at that time of day.

Success criteria

Ask the learners:

- What time does this analogue clock show? Say the same time in a different way.
- What time does this digital clock show? Say the same time in a different way.
- Can you show twenty to three on an analogue clock?
- What does half past six look like on a digital clock?

Ideas for differentiation

Support: For the starter activity, group these learners with a more confident partner.

Extension: In the final main activity, challenge these learners to include some times shown on analogue clocks to the nearest minute.

Time ordering cards

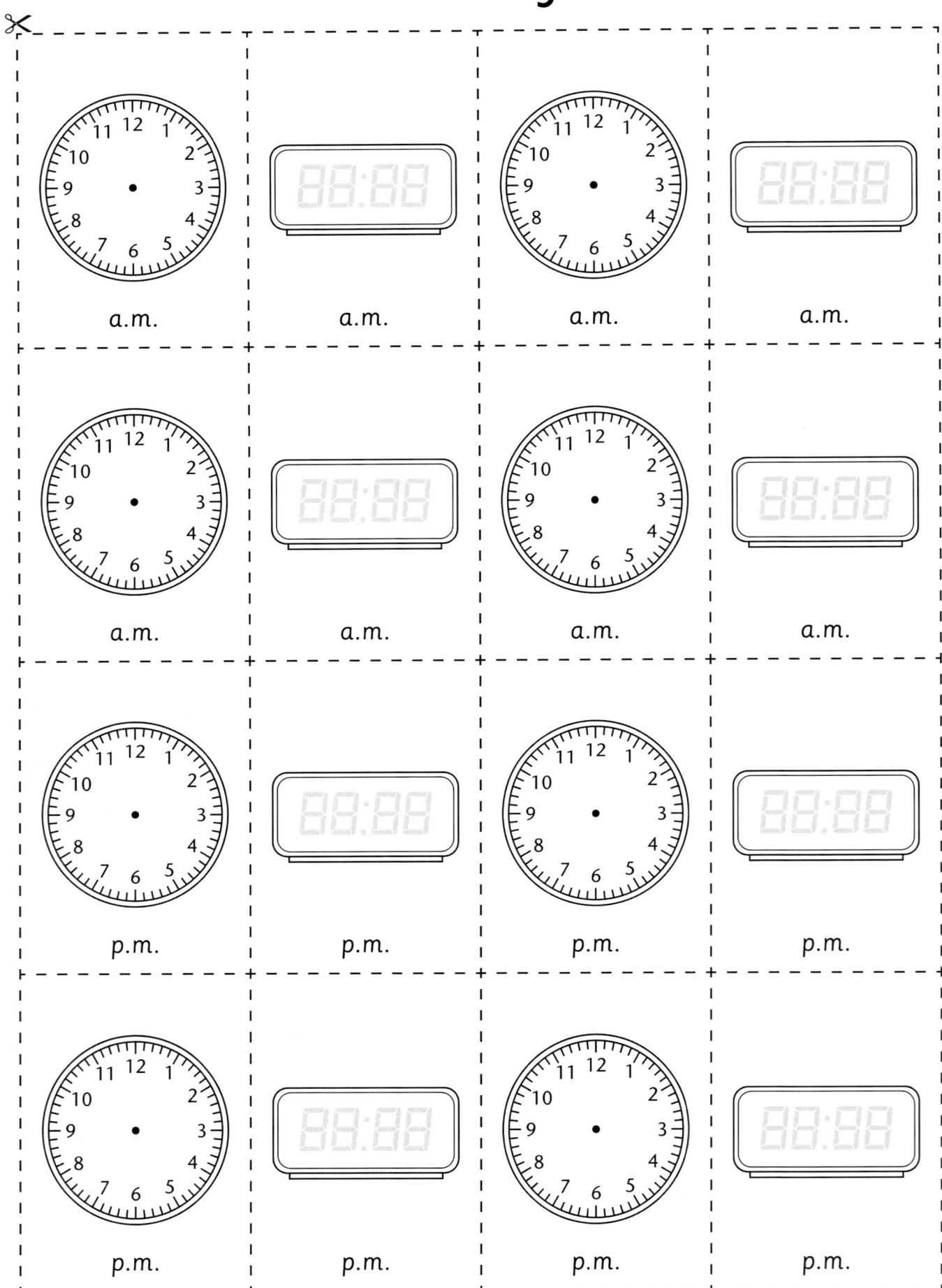

 Cambridge Primary: Ready to Go Lessons for Maths Stage 3 © Hodder & Stoughton Ltd 2013

Time 9

Learning objectives

- Solve word problems involving measures. (3Ml5)
- Begin to calculate simple time intervals in hours and minutes. (3Mt3)
- Estimate and approximate when calculating, and check working. (3Pt10)

Resources

Counting stick; photocopiable page 180; clock faces with moveable hands; calendars.

Starter

- Count using a counting stick. Point to each division in turn, counting aloud, and encouraging the learners to count aloud with you:
  - Count on in whole hours from any time (for example: *Fourteen minutes past three, fourteen minutes past four, fourteen minutes past five, fourteen minutes past six …*).
  - Count on in intervals of five minutes from times given to the nearest five minutes (for example: *Ten fifty, ten fifty-five, eleven, eleven-oh-five …*).
  - Count on in intervals of one minute from any time (for example: *Nine forty-eight, nine forty-nine, nine fifty, nine fifty-one …*).

Main activities

- Display a copy of photocopiable page 180, and read out one of the word problems.
- Ask the learners to estimate the answer, and describe the strategies they used.
- Work through the problem, asking the learners to suggest the method. Use a clock face and a calendar to support calculation where appropriate.
- Ask the learners to compare the answer with the estimate. Ask: *Given the estimate, does the answer seem reasonable? What would it mean if the estimate and the answer were very different?* (It might mean the answer is wrong.) *What else could you do to check the answer?* (You could repeat the calculation using a different method.)

- Work through another word problem in the same way.
- Give each learner photocopiable page 180. Ask the learners to complete it, working either individually or in pairs. Remind them to make an estimate for each question and use it to check the reasonableness of their answer. Ensure the learners have access to clock faces and calendars.

Plenary

- Ask the learners to give the answers to the word problems on photocopiable page 180, and describe the methods they used.
- Ask volunteers to read out the word problems they wrote. Ask the learners to solve one of the problems.

Success criteria

Ask the learners:

- What did you need to do to solve this problem?
- What was your estimate? Explain how you worked it out.
- What answer did you get? Explain how you worked it out.
- Is your answer reasonable? How can you tell?

Ideas for differentiation

Support: Group these learners together and guide them through an extra problem. Ask them to complete only the first four problems, as these are the easiest.

Extension: Ask these learners to make up word problems based on their own ideas instead of the questions given on photocopiable page 180.

Time problems 3

1. A boat trip starts at 3.20p.m. and lasts 2 hours 45 minutes. When does the boat trip end?

   _____

2. Azeem's three-week holiday ends on 25th August. What date does it start?

   _____

3. A film starts at 7.35p.m. and lasts 1 hour 45 minutes. What time does the film finish?

   _____

4. Haruto's birthday is on 13th November and his brother Kaito's birthday is on 6th December. On Haruto's birthday, how long does Kaito have to wait for his birthday?

   _____

5. Hassan started a race at 11.15a.m. and finished it at 12.45p.m. How long did it take Hassan to run the race?

   _____

6. Write a word problem to go with each of these questions:

   a) How much later than 10.15a.m. is 1.20p.m.?

   _____

   _____

   b) What is the date 20 days later than 15th October?

   _____

   _____

   c) What is the time 2 hours and 25 minutes earlier than 5.15p.m.?

   _____

   _____

Money 5

- Consolidate using money notation. (3Mm1)
- Use addition and subtraction facts with a total of 100 to find change. (3Mm2)
- Explain a choice of calculation strategy and show how the answer was worked out. (3Ps2)

Pencils; plain paper; photocopiable page 182; lots of price-labelled items, priced to the nearest cent (all prices under $10); US coins; $1 bills.

Starter

- Hand out paper and pencils.
- On the board write a multiple of 5 between 5 and 95. Ask: *How many more to make 100?* and ask the learners to write the answer. Repeat, in a random order, for all the multiples of 5 between 5 and 95.
- Repeat the activity for whole numbers less than 100 that are not multiples of 5 (for example 28, 73, 56, 47, 32 ...). Ask various learners to explain how they worked out the answer. Encourage every learner to try out several different strategies.

Main activities

- Display a copy of photocopiable page 182. Use it to model finding change from $1. Make explicit the link with the work done on finding complements to 100 in the starter activity ($1 is 100 cents).
- Distribute the price-labelled items, coins and $1 bills. Ask the learners to work in pairs to sell each other items. The buyer must offer the seller $1 and the seller make and give the correct change. Both learners should record the transaction.

Plenary

- Display two priced items. Ask: *If you bought both of these, how much change would you get from $1?* Allow the learners to work out the answer with a partner.

Ask the learners:

- An item costs 58c and you pay for it with $1. How much change will you get?
- How did you work out the answer?
- Why did you decide to work out the answer that way?
- If you get 35c change from $1, how much did you spend?

Support: For the final main activity, pair these learners together. Give them items priced in multiples of 5c.

Extension: In the final main activity, pair these learners together and ask them to buy two items at a time.

Name: _____

How much change?

Look at how much each item costs and then work out how much change should be given.

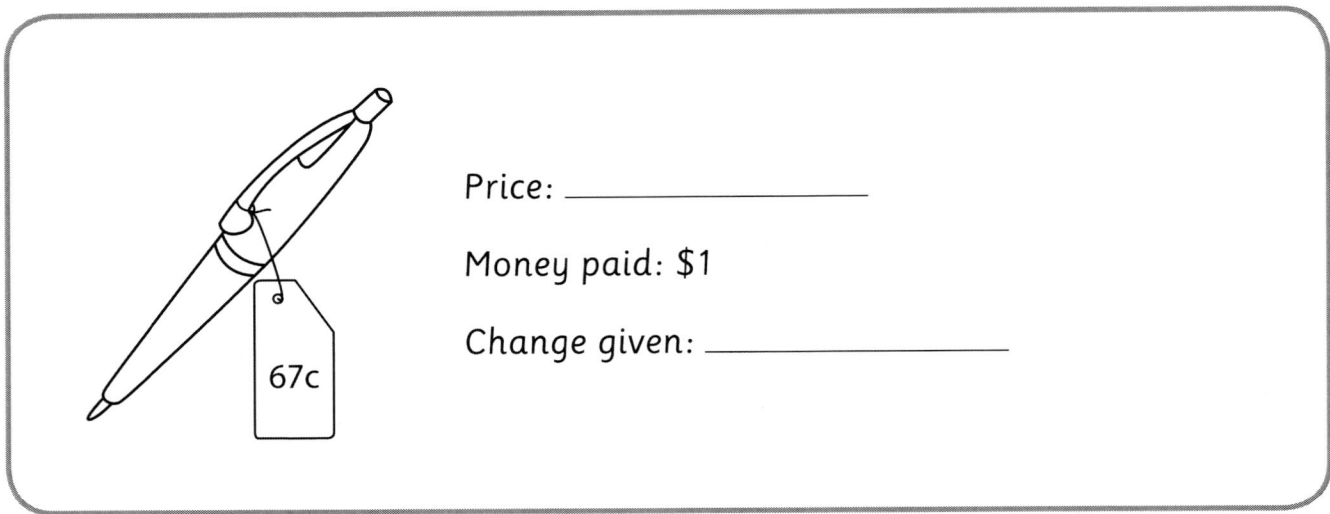

Price: _____

Money paid: $1

Change given: _____

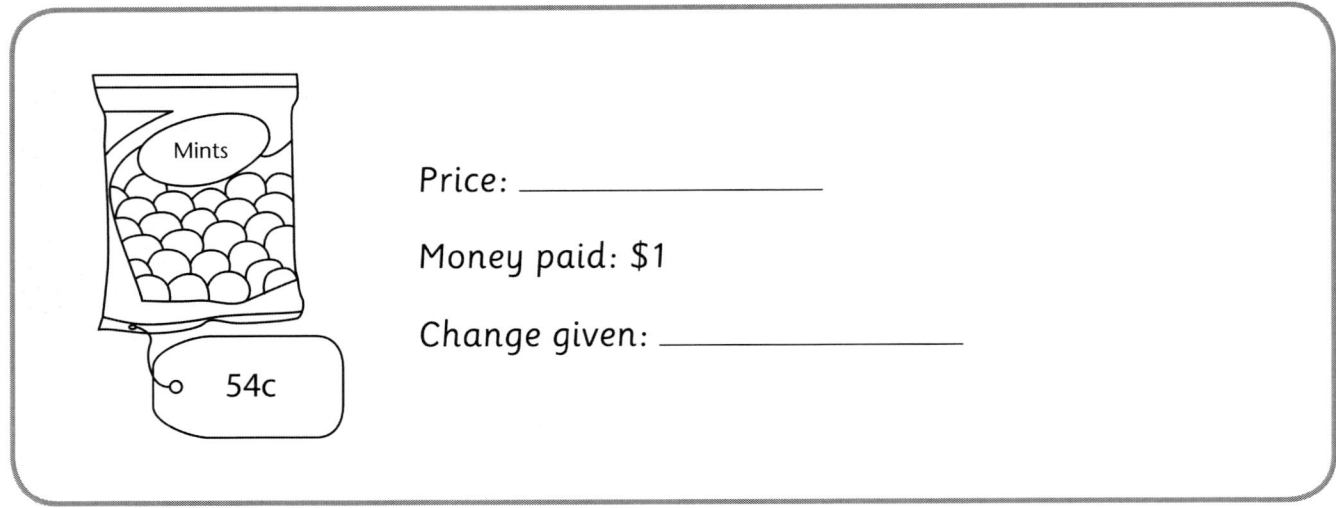

Price: _____

Money paid: $1

Change given: _____

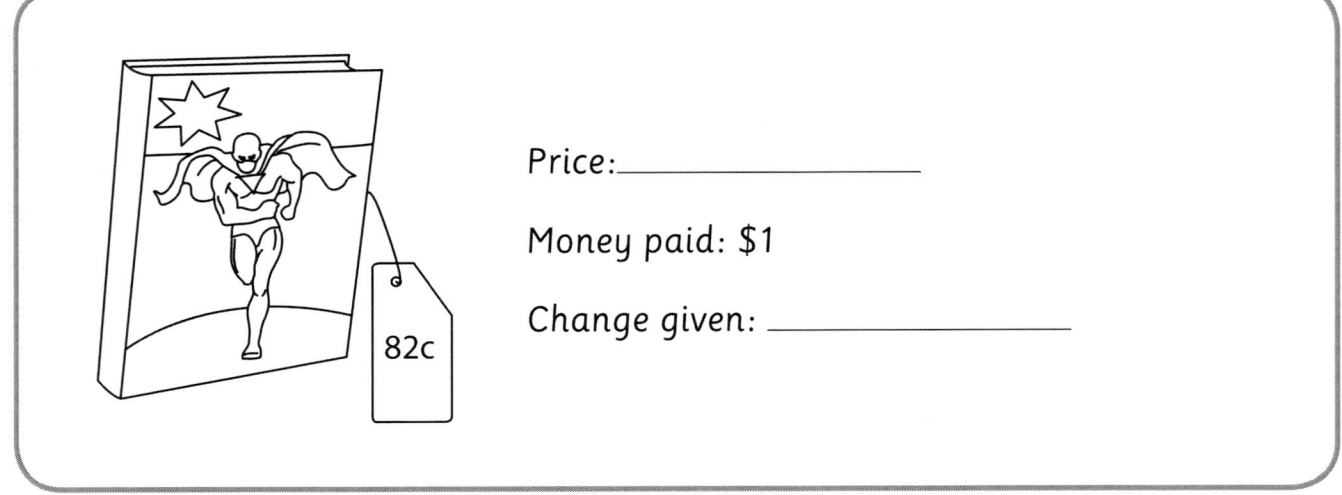

Price: _____

Money paid: $1

Change given: _____

Cambridge Primary: Ready to Go Lessons for Maths Stage 3 © Hodder & Stoughton Ltd 2013

Money 6

Learning objectives

- Solve word problems involving measures. (3Ml5)
- Choose appropriate mental strategies to carry out calculations. (3Pt1)
- Consider whether an answer is reasonable. (3Pt12)
- Make up a number story to go with a calculation, including in the context of money. (3Ps1)

Resources

A set of large arithmetical facts cards (see Starter); 0 to 9 number fans (see photocopiable page 24); photocopiable page 184.

Starter

- Before the lesson make a set of large arithmetical facts cards with multiplications and divisions from the 2, 3, 4, 5 and 10 times tables, and additions and subtractions to 20 (for example 6×5; $20 \div 4$; $9 + 8$; $20 - 12$, and so on). Ensure that none of the answers has two digits that are the same.
- Give each learner a 0 to 9 number fan. Hold up the cards one at a time, asking the learners to show the answer on their number fans. Keep the pace as brisk as possible.

Main activities

- Give the learners four money calculations (with totals up to $1): one addition, one subtraction, one multiplication and one division (no remainders), for example $29c + 63c$; $95c - 46c$; $3 \times 31c$; $75c \div 5$. For each calculation, ask the learners to choose their own strategy, and then describe how they worked out the answer. Discuss the variety of approaches and how to decide whether an answer is reasonable.
- Display a copy of photocopiable page 184 and read through the problems together. For each problem, ask the learners to identify the calculation they will need to do, and give their initial thoughts about possible answers.

- Organise the learners into pairs and give each pair photocopiable page 184. Ask the learners to work through the problems, keeping any jottings they make so that they can remember the strategies they used to solve each problem.

Plenary

- Ask the learners to give the answers to the first six problems, explaining how they worked out each answer, and why they chose that particular strategy.
- Discuss question 7, asking the learners to explain how they worked out the answer. Collect all answers found, asking: *Have we found all the answers? How can we check?* (There are seven possible ways of spending $2.40 by buying apples at 20c and bananas at 40c: 12A 0B; 10A 1B; 8A 2B; 6A 3B; 4A 4B; 2A 5B; 0A 6B.)

Success criteria

Ask the learners:

- Choose a word problem that you have solved. What calculation or calculations did you need to do in order to solve it?
- What is the answer to the problem?
- How did you work it out?

Ideas for differentiation

Support: Group these learners together and guide them through one or two of the problems. Ask them to complete only the first four problems, as these are the easiest.

Extension: Ask these learners to work on photocopiable page 184 individually, rather than in pairs.

Name: _____

Money problems

1. Ben buys a pack of stickers for 74c.
   How much change should he get from $1?

   _____

2. Momoka buys a carton of milk for 58c and a loaf of bread for 39c.
   How much does she spend altogether?

   _____

3. Dad spends 50c every day on a newspaper.
   How much does Dad spend on newspapers every week?

   _____

4. Farida spends 84c buying four identical packets of biscuits.
   How much does each packet of biscuits cost?

   _____

5. Satnam buys a ruler for 25c, a pencil case for 19c and a pen for 45c.
   How much does he spend altogether?

   _____

6. Annika buys some sweets for 49c and some itching powder for 38c.
   How much change should she get from $1?

   _____

7. Apples are 20c each and bananas are 40c each.
   You have $2.40. How many apples and how
   many bananas could you buy?

   _____

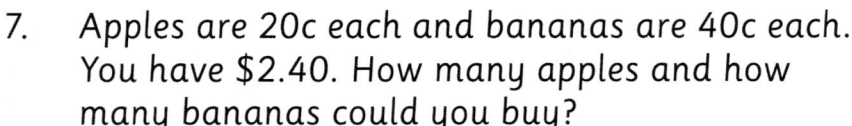

 Cambridge Primary: Ready to Go Lessons for Maths Stage 3 © Hodder & Stoughton Ltd 2013

Length 3

Learning objectives

- Choose and use appropriate units and equipment to estimate, measure and record measurements. (3Ml1)
- Know the relationship between kilometres and metres, metres and centimetres, kilograms and grams, litres and millilitres. (3Ml2)
- Use a ruler to draw and measure lines to the nearest centimetre. (3Ml4)

Resources

Cards made from photocopiable page 186; timer; metre stick; plan of your classroom; plain paper; rulers.

Starter

- Revise the relationships between various units of measure by playing a follow-me card game with the cards made by enlarging photocopiable page 186 onto A3 card and cutting out. Find and keep the START card. Give out the rest of the cards. Read the START card aloud. The learner who has the answer on their card should read their card aloud, and so on, until the END card is reached.
- Play a few more rounds, redistributing the cards between rounds. Time the activity, challenging the learners to keep improving on their time.

Main activities

- Ask the learners to estimate a few dimensions in the classroom (for example the height of a desk, the length of the board) by comparing them to the length of a metre stick. (At least one of the dimensions should be greater than 1 metre.) Record the estimates in centimetres or in metres and centimetres.
- Demonstrate using a metre stick to measure to the nearest centimetre the dimensions you have estimated. Record each measurement in centimetres if less than 1 metre and in metres and centimetres if greater than 1 metre. Compare the measurements to the estimates.

- Organise the learners into small groups of two or three, giving each group a metre stick and a plan of the classroom. Ask the learners to estimate and measure the dimensions of the classroom, writing estimates and measurements on the plan. Challenge the learners to make accurate measurements, and to improve the closeness of their estimates as they work.

Plenary

- On the board write six lengths up to 30 cm. The learners should draw a line of each length on plain paper with a ruler, then measure each other's lines to check they are drawn correctly.

Success criteria

Ask the learners:

- What is the relationship between kilometres and metres / kilograms and grams / litres and millilitres?
- How many centimetres are there in half a metre?
- Name something in the classroom that is over 1 metre long. About how long do you think it is? Measure and record its length.

Ideas for differentiation

Support: In the final main activity, group these learners with learners who will support them and encourage them to participate.

Extension: Ask these learners to draw a sketch map of another, larger room in the school (for example the hall), estimate and measure its dimensions, and write these on the map.

Measure follow-me cards

START Who has half a kilometre?	I have 500 metres. Who has 1 kilogram?	I have 1000 grams. Who has a quarter of a litre?	I have 250 millilitres. Who has 1 metre?
I have 100 centimetres. Who has half a metre?	I have 50 centimetres. Who has half a kilogram?	I have 500 grams. Who has 2 metres?	I have 200 centimetres. Who has 2 kilograms?
I have 2000 grams. Who has one tenth of a metre?	I have 10 centimetres. Who has 5 kilometres?	I have 5000 metres. Who has 2 litres?	I have 2000 millilitres. Who has half a litre?
I have 500 millilitres. Who has one tenth of a kilometre?	I have 100 metres. Who has 3 litres?	I have 3000 millilitres. Who has a quarter of a kilometre?	I have 250 metres. Who has 4 metres?
I have 400 centimetres. Who has a quarter of a metre?	I have 25 centimetres. Who has 2 kilometres?	I have 2000 metres. Who has a quarter of a kilogram?	I have 250 grams. Who has one tenth of a litre?
I have 100 millilitres. Who has 4 kilograms?	I have 4000 grams. Who has 3 metres?	I have 300 centimetres. Who has 4 kilometres?	I have 4000 metres. **END**

 Cambridge Primary: Ready to Go Lessons for Maths Stage 3 © Hodder & Stoughton Ltd 2013

Mass 3

Learning objectives

● Choose and use appropriate units and equipment to estimate, measure and record measurements. (3MI1)

● Read to the nearest division or half division, use scales that are numbered or partially numbered. (3MI3)

● Begin to understand everyday systems of measurement in length, weight, capacity, time and use these to make measurements as appropriate. (3Pt2)

Resources

Photocopiable page 188; objects with a variety of masses; 1 kg masses; 100 g masses; sticky notes; variety of mass-measuring instruments (e.g. pan balances, spring scales, kitchen scales and bathroom scales).

Starter

• Show the cards from an enlarged copy of photocopiable page 188 one at a time. Ask: *Which unit would you use to measure the mass of this object – grams or kilograms?*

• Display all the cards made from photocopiable page 188 together, asking the learners to order the objects according to their mass.

Main activities

• Display four objects, two with masses under 1 kg (for example a book and a football) and two with masses over 1 kg (for example a 2-litre carton of milk and a bag of flour).

• Ask a volunteer to directly compare the mass of each object with a mass of 1 kg, and sort the objects into two groups: < 1 kg and > 1 kg.

• Ask two volunteers to compare the masses of the objects in each group, and use sticky notes to label them 'lighter' and 'heavier'.

• Ask another two volunteers to estimate the mass of each object to the nearest 100 grams. (Give them 1 kg and 100 g masses for direct comparison.) Ask the volunteers to label each object with its estimated mass.

• Ask another two volunteers to choose and use appropriate instruments to measure the mass of each object. Record the mass of each object as accurately as the measuring instrument will allow.

• Organise the learners into groups of between four and six. Give each group a variety of about a dozen objects, some with masses less than 1 kg and some with masses more than 1 kg. Ask the learners to use sorting and direct comparison to estimate the mass of each object, and then measure the mass of each object using an appropriate measuring instrument.

Plenary

• Ask: *How could you check a mass measurement?* (For example compare it with someone else's; compare it to your estimate; re-measure.)

• Write down six masses over 1 kg, some in grams, and some in kilograms and grams. Ask the learners to order the masses.

Success criteria

Ask the learners:

● Do you think the mass of a dictionary is less than or greater than 1 kg?

● Estimate the mass of a dictionary. How did you work out your estimate?

● What equipment could you use to measure the mass of a dictionary?

● What is the mass of a dictionary? How close was your estimate?

Ideas for differentiation

Support: For the final main activity, group these learners together and work with them.

Extension: Ask these learners to choose extra objects from around the classroom, and estimate and measure their masses.

Which unit of mass?

the mass of a pencil

the mass of a baby

the mass of an exercise book

the mass of a fridge

the mass of an apple

the mass of an elephant

the mass of a plate

the mass of a bus

 Cambridge Primary: Ready to Go Lessons for Maths Stage 3 © Hodder & Stoughton Ltd 2013

Capacity 3

Learning objectives

- Know the relationship between kilometres and metres, metres and centimetres, kilograms and grams, litres and millilitres. (3Ml2)
- Solve word problems involving measures. (3Ml5)

Resources

Follow-me cards made from photocopiable page 186; timer; photocopiable page 190.

Starter

- Revise the relationships between various units of measure by playing a follow-me card game with the cards made from photocopiable page 186. Find and keep the START card. Give out the rest of the cards. Read the START card aloud. The learner who has the answer on their card should read their card aloud, and so on, until the END card is reached.
- Play a few more rounds, redistributing the cards between rounds. Time the activity, challenging the learners to beat their previous best time.

Main activities

- Give the learners four calculations involving capacity: one addition, one subtraction, one multiplication and one division (no remainders), for example 350 ml + 175 ml; 92 litres – 48 litres; 5 × 23 litres; 1 litre 400 ml ÷ 4. For each calculation, ask the learners to choose their own strategy, and then describe how they worked out the answer. Discuss the variety of approaches.
- Display a copy of photocopiable page 190 and read through the problems together. For each problem, ask the learners to identify the calculation they will need to do, and give their initial thoughts about possible answers.
- Organise the learners into pairs and give each pair a copy of photocopiable page 190. Ask the learners to work through the problems, keeping any jottings they make so that they can remember the strategies they used to solve each problem.

Plenary

- Ask the learners to give the answers to the problems on photocopiable page 190, explaining how they worked out each answer.
- Discuss various ways of expressing some answers (for example 2500 ml can also be expressed as 2 litres 500 ml or as $2\frac{1}{2}$ litres.)

Success criteria

Ask the learners:

- Choose a word problem that you have solved. What calculation or calculations did you need to do in order to solve it?
- What is the answer to the problem?
- Can you express the answer in a different way?
- How did you work out the answer?

Ideas for differentiation

Support: Group these learners together and guide them through one or two of the problems. Ask them to complete only the first four problems, as these are the easiest.

Extension: Ask these learners to write their own word problems, and give them to a friend to solve.

Name: _____

Capacity problems

1. A recipe uses 375 ml of water and 440 ml of milk.
   How much liquid is that altogether?

   _____

2. A one-cup measure holds 250 ml. A quarter-cup measure holds 62 ml.
   What is the difference in the capacity of the two measures in millilitres?

   _____

3. Josef has three bottles of lime juice. Each bottle contains 230 ml.
   How many millilitres of lime juice does Josef have?

   _____

4. A jug contains 850 ml of orange juice. Raul pours 325 ml of the orange
   juice into a glass. How much orange juice is left in the jug?

   _____

5. Tito makes 1 litre of soup. He wants to divide it equally between five people.
   How much soup should he give to each person?

   _____

6. To fill a fish tank, Abdul pours in five 10-litre buckets of water and
   Jameel pours in seven 5-litre buckets of water.
   How many litres does the fish tank hold?

   _____

7. The petrol tank in Akim's car holds 56 litres, but at the moment it is empty.
   Akim has three petrol cans. One contains 8 litres of petrol, one contains
   12 litres and one contains 5 litres.

   a) How much petrol does Akim have? _____

   b) How much more petrol does Akim need in order to fill his car's tank?

   _____

Cambridge Primary: Ready to Go Lessons for Maths Stage 3 © Hodder & Stoughton Ltd 2013

Unit assessment

● What units of time do you know? How are they related?

● What units would you use to measure the height of a person? What measuring instrument would you use?

● What change would you get from $5 if you spent $3.72? How did you work out the answer?

● How many millilitres are there in a quarter of a litre? Name a container that holds about a quarter of a litre.

Summative assessment activities

Observe the learners while they take part in these activities. You will quickly be able to identify those who appear to be confident and those who may need additional support.

Measures matching game

This game assesses the learners' knowledge of the relationships between kilometres and metres, metres and centimetres, kilograms and grams, and litres and millilitres.

You will need:

Cards made from photocopiable page 192.

What to do

• Organise the learners into groups of between four and six. Give each group a set of cards made from enlarging photocopiable page 192 onto A3 card. One learner should shuffle the cards and place them face down in a 6 by 6 grid pattern.

• Players should take it in turns to turn over two cards. Everyone must be able to see the two turned-over cards and their position. If the two cards show a matching measurement (for example 2 km and 2000 m), the player who turned them over keeps both cards. If the two cards do not match, the player turns them back over.

• The game is over when there are no more cards left. The winner is the player with the most cards.

Calendar activity

This activity assesses the learners' ability to read a calendar and calculate time intervals in weeks and days.

You will need:

Calendars.

What to do

• Organise the learners into pairs, giving each pair a calendar for the next two months. Ask the learners a few problems that require them to calculate time intervals in weeks and days, for example ask: *How long is it from … to …? What is the date x weeks and x days before / after …? If I go on holiday on … and spend x weeks and x days away, what date does my holiday end?*

• Ask each pair to write five (or more) similar questions to do with intervals of time on the calendar. Ask pairs to write a separate answer sheet for their questions, and then give their questions to another pair to answer.

Give each learner a copy of photocopiable page 178 on which you have filled in 16 different times (5-minute intervals only). Leave the page in one piece. Do not cut the cards out.

Ask the learners to cut out the cards, and then order the times from earliest to latest.

July						
M	T	W	Th	F	Sa	Su
			1	2	3	4

Measures matching cards

5 km	5000 m	1 km	1000 m
$\frac{1}{10}$ km	100 m	$\frac{1}{2}$ km	500 m
5 m	500 cm	$\frac{1}{2}$ m	50 cm
4 m	400 cm	$\frac{1}{4}$ m	25 cm
$\frac{1}{2}$ kg	500 g	2 kg	2000 g
$1\frac{1}{2}$ kg	1500 g	$\frac{1}{4}$ kg	250 g
4 kg	4000 g	1 litre	1000 ml
$\frac{1}{2}$ litre	500 ml	5 litres	5000 ml
$\frac{1}{10}$ litre	100 ml	$\frac{1}{4}$ litre	250 ml

Cambridge Primary: Ready to Go Lessons for Maths Stage 3 © Hodder & Stoughton Ltd 2013